ROLLING AND HIGH PLAINS OF TEXAS AND SOUTHWESTERN OKLAHOMA

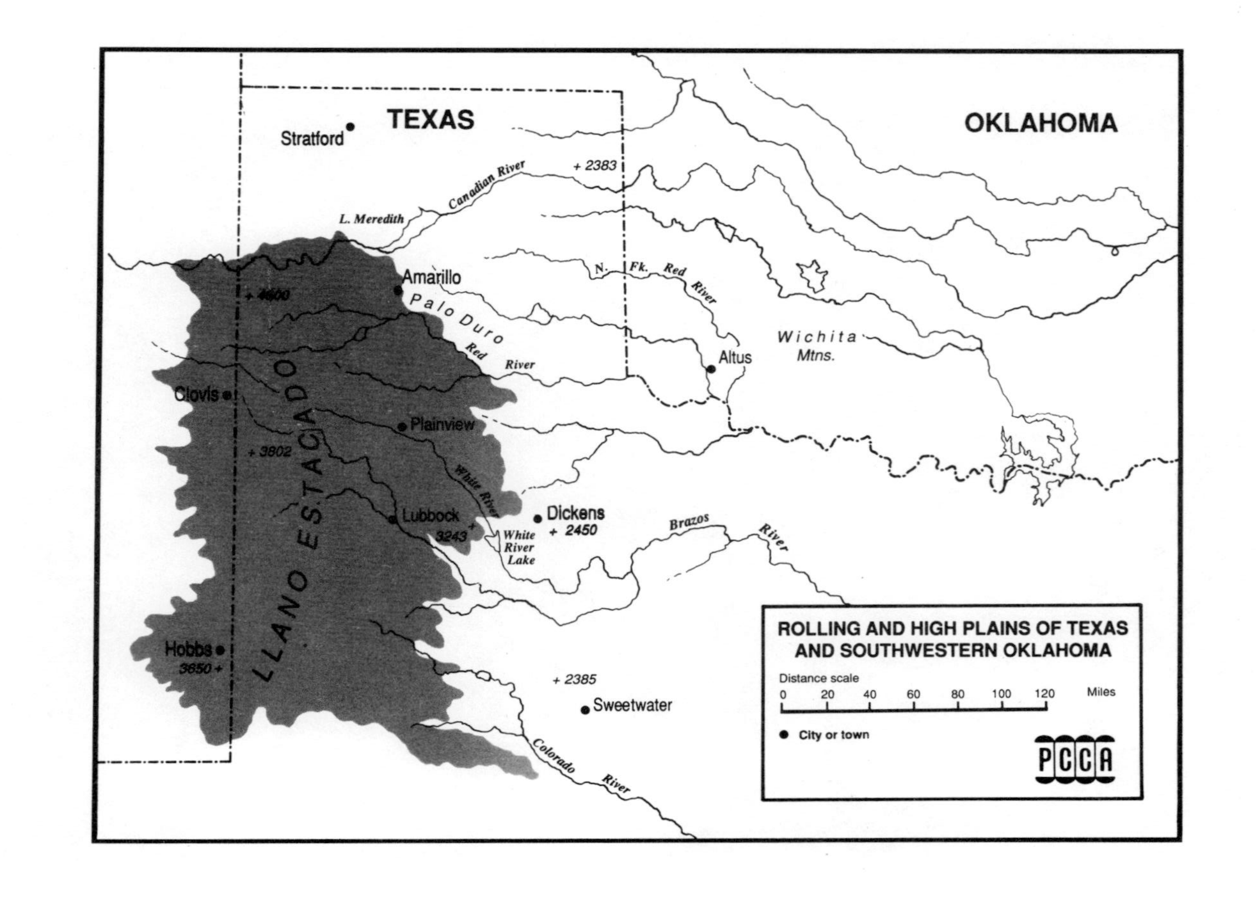
TEXAS
OKLAHOMA
Stratford
Canadian River
+ 2383
L. Meredith
Amarillo
+ 4600
Palo Duro
Red River
N. Fk. Red River
Wichita Mtns.
Altus
Clovis
LLANO ESTACADO
Plainview
+ 3802
White River
Lubbock
3243 +
Dickens
+ 2450
White River Lake
Brazos River
Hobbs
3650 +
+ 2385
Sweetwater
Colorado River
ROLLING AND HIGH PLAINS OF TEXAS AND SOUTHWESTERN OKLAHOMA
Distance scale
0 20 40 60 80 100 120 Miles
City or town
PCCA

Field to Fabric

AMERICAN

cotton

GROWERS

Field to Fabric

THE STORY OF AMERICAN COTTON GROWERS

By Jack Lichtenstein

TEXAS TECH UNIVERSITY PRESS

LUBBOCK · 1990

This book was set in 10 on 12 Baskerville and printed on acid-free paper that meets the guidelines for permanence and durability of the Committee on Production Guidelines for Book Longevity of the Council on Library Resources.

Printed in the United States of America

Library of Congress Cataloging-in-Publication Data

Lichtenstein, Jack.
Field to fabric : the story of American Cotton Growers / Jack Lichtenstein.
p. cm.
Includes index.
ISBN 0-89672-238-4 (cloth)
1. American Cotton Growers (Firm)--History. 2. Cotton trade--United States--History. 3. Cotton textile industry--United States--History. I. Title.
HD9079.A44L53 1990
334'.687721'0973--dc20 90-37181
CIP

Texas Tech University Presss
Lubbock, Texas USA 79409-1037

Contents

Field to Fabric

Introduction

This is the story of an American enterprise. It was called American Cotton Growers (ACG), and it was a West Texas agricultural cooperative. It lived less than 17 years, from early in 1973 until September of 1989. Then it passed into history. Seventeen years is a short time as successful enterprises go, and this was a very successful enterprise. But as the act of picking up the book will attest, this is not a short story. The 17 years of ACG were filled with people and events that surged through their time and place like storm waters down a canyon riverbed.

Why write a history of a West Texas cotton cooperative? One reason that should be approached head-on is that American Cotton Growers retained the author to write their story. Cynics may think this self-serving, but it is not at all unusual for an organization to retain someone to write its history, particularly with good reason to see the history as special or noteworthy. When the organization is a cooperative comprised of thousands of owner-members who have invested their livelihoods in it, as was ACG, the case is all the more compelling. They have a right to the pleasure and pain of memories. They also have a right to facts that may have escaped their notice and points of view they may not have heard before, whether they reinforce or refute their own.

Those who might criticize a business for contracting to have its own history written should note that failed enterprises do not write histories. If an organization is successful, the decisions and actions of those responsible should be recorded for posterity. Thousands of enterprises fail every year and slip beneath the waves of American business, most leaving barely a ripple and certainly not a study of what was done, and how, and by whom. One can only imagine how much their former owners, members, or employees would like to ''curl up with a book'' telling the story of how they succeeded beyond their wildest dreams.

Being retained to write a history can be a touchy proposition. The goal of the author is to portray persons and events as nearly accurately as possible. Despite the client's patronage, the author wants to maintain professional integrity and does not want to mislead or misinform. On the other hand, the author has to be somewhat sensitive to how the portrayal of events and persons will affect the client, especially where the history is recent and the participants still in close contact with one another. No organization in its right institutional mind is going to pay to have itself killed in print. The author retained to write a history must hope there is nothing in that history that will put him in the position of having to decide between accuracy and integrity. The author of this book fared well in that regard. Drafts of the manuscript were submitted to a small number of ACG managers for review. There were several issues on which the author and ACG management disagreed. In most cases, the disagreement was about a choice of words, not about whether an issue should or should not be included or about the basic thrust of a portrayal.

Where there were points of contention between the author and the client, a solution would be reached after vigorous debate. In the author's opinion, the solutions reached addressed the concerns of client and author without compromising either accuracy or integrity.

Participants in the history who were interviewed were given an opportunity to express their recollections however they wanted. A person's recollections and points of view are, after all, that person's property, and who are we to say that he or she is not remembering or interpreting events correctly? If enough people are interviewed and their perspectives recorded, an accurate picture emerges. The author was willing to rewrite portions of the manuscript to allow for changes in interpretation, but not if it seemed to alter the nature of events. Nor was he asked to.

When it came time to decide whether a controversial issue should be included in the history, the decision was solely the author's, and he applied two tests. First, did the issue materially affect the history of American Cotton Growers? The book is not without ''sideshows,'' or incidental events; there are hundreds. But there are no controversial events slapped on for the sake of gratuitous controversy. Second, would discussion of the issue in-vade the privacy of an individual or organization by airing the details of agreements that were legal but of a confidential nature?

American Cotton Growers was a private, not public, entity. It was entitled to a certain amount of privacy in conducting its affairs, until such time as its member-owners would vote otherwise. And they did not. If the answer to the first question was no, or the second yes, the issue is not in here. It is that simple.

The author tried hard to find both heroes and villains in the story of ACG. Heroes abound, and the villains seem guilty of no more than common errors that helped place the organization in its frequent troubles but were offset by contributions that helped lead to its success. There is plenty of misfeasance, but little malfeasance. In other words, the history of American Cotton Growers is more like life than myth. This is not to say the story is without its moments of irony, excitement, exhilaration, despair, and intrigue.

Admittedly, the history of a cotton cooperative sounds like a sure cure for insomnia. Business histories normally don't make the most interesting reading, other than to a few narrow segments of society including those who genuinely enjoy reading them, those who are seeking lessons that might be applied in other settings, and those who are mentioned in the book (and their mothers). Business-history aficionados may find reading about a cooperative instructive, or at least an interesting change of genre.

The book is written, first and foremost, for the enlightenment and enjoyment of the people who lived the story. The story is one of people — their relationships, decisions, and actions. "People story" is a hackneyed expression, but it happens to fit here. It is enough to chronicle a "bunch of West Texas cotton farmers" who decided to go into the denim business and built what is arguably the world's best and most profitable denim mill. To break that story into pieces in order to show developments in minute detail would be unforgivable. Nevertheless, the numbers that really matter are here.

Those seeking lessons from this book may come away somewhat disappointed. Because of the unique structure of the cooperative, whole portions of the experience simply don't apply to other types of organizations. Furthermore, much of what happened to ACG in its history, good or bad, was totally out of the cooperative's control. It is doubtful whether there ever has been a greater confluence of talent, ability, ambition, grit, and pure "dumb luck" anywhere in cooperative agriculture. The luck was not all good. But, on balance, it was phenomenal.

The book does have lessons. Some very basic principles are woven into ACG's evolution, resulting in a historical fabric as strong as ACG denim. The principles may seem so obvious that the recitation of them is pedantic. However, they are routinely ignored or violated by organizations everywhere. Among them: different stages in an organization's history may call for different types of leaders; hire the best people, stand back and let them do their jobs; seek and cultivate strong allies; business by handshake and mutual respect is preferable to business by contract and the power of enforcement, but where contracts are necessary, they should be enforced; never underestimate the power of confidence, either in convincing yourself or moving others; pyrrhic victories are foolish, compromise is preferable to conflict, particularly where the parties are closely related.

Much of ACG's history was lived through one state of crisis or another. There are many points of view about the crises, how they evolved, who was responsible, how they should have been resolved, and so forth. It is difficult today to come to definitive conclusions about these events. There are two reasons for this.

First, American Cotton Growers never can be faulted for putting too much in writing. The seven-member Board of Directors and the full-time staff had procedures for conducting their duties, but they seemed principally to ensure adherence to laws, rules, parliamentary order, and mutual respect. They were partly formal and partly based on understanding between gentlemen of like background and goals. The procedures and records were not aimed at establishing and perpetuating a bureaucracy. Nor were they structured in such a way that information would become public that might violate confidences or jeopardize business relationships. Because the success of the cooperative would rely mainly on the relationship with one customer, Levi Strauss & Co., they always feared that proprietary information somehow would make its way into the public domain and damage or destroy the arrangement. It was a solid arrangement, based on hard work and trust and the ACG Board did not want it undermined.

A cooperative is very much more like a family than other types of business enterprises, particularly in agriculture. The members commit a great deal to the cooperative and depend upon it for a great deal in return. There is pressure on the board and management to provide information to them. In interviews, the author talked with several members, particularly young farmers, who felt that too much of what the board and management did was done in secrecy. One of the best examples of the secretive nature of ACG was when the members of its own textile pool committee made an early 1986 request for printed reports of ACG deliberations to take back to their gins and share with their members. The ACG Board answered the request with paper, but not in the form of reports. The pool committee members were given notepads and pencils at future meetings.

There is an uncertain middle ground between information that is truly proprietary — which should be protected in the best interests of the entire cooperative family — and information that is needed by members to assure them business is being conducted properly. The issue of anyone's "right to know" will not be settled in this book. If one judges by results, the board and management may be considered to have conducted their duties very well on behalf of the members. Whether things would have gone better or worse if the members had more information is impossible to answer. The ACG leadership felt it let out all the information it could. Whether it did was a matter for the membership to decide. The cooperative had its own ways of operating, and they were not conducive to writing a history.

The author had access to all official records of ACG, but these were not much to go on. The board meeting minutes of nearly 17 years provide a chronological framework, essential for putting events in their proper order, but little in the way of detail. In most cases, they tell *what* happened, albeit briefly. In some cases, they tell *why* things happened; by looking for cause-and-effect relationships between events, one often can determine the "why." But rarely do the records tell *how* things happened.

As meeting minutes are wont to do, they lack description. The most significant or incredible incidents are reported as clinically as the calling of the meeting to order. That is no one's fault. It is simply the way minutes are. There also seems to have been an especially careful effort in ACG to minimize friction and the effects of controversy. There are hints of conflict in the minutes, but there is precious little about the give and take of ideas and opinions that led the board to do what it did. Sometimes there is an oblique reference to a "lively discussion of the issue," then the result is reported with little or nothing about the process through which it was derived. In addition to the board-meeting minutes, the author found articles from publications, most of them internal publications, that provide only some detail.

With a dearth of written materials from which to work, the author depended primarily on interviews for details. For this reason, the book is filled with the personal reminiscences of those people who played a part in the story. More than fifty interviews were conducted, ranging from multiple, tape-recorded sessions with some key figures to casual conversations with others. Given a choice between building a history from reams of paper or personal interviews, this author will opt every time for the latter (despite having just complained about ACG's lack of written records). In this case, the interviews were the only meaningful medium with which to work, so there was no choice. A history described by eyewitnesses is much more interesting than one described by official records. Interviews have a basic weakness in that the history emanating from them is strained through a sieve of personal perspective, which might include bias, self-service, revenge, guilt, and ambition. In its own way, however, it is more accurate than that emanating from papers in that it accounts for human attitudes, perceptions, and beliefs.

The interview process leads to the second reason why it was difficult to determine what happened and how. West Texas cotton farmers and those who work with them tend toward taciturnity, gentility, and self-effacement. When they are brash or aggressive, it is with regard to the actions they take rather than the people with whom they deal. It is very difficult to get them to "open up" in an interview, to elaborate on their own roles, or to be critical of someone else's. In an age when so many seem willing to do anything to enhance themselves or bring others low, their bent is in exactly the opposite direction.

Asked about the most turbulent, controversial incidents in ACG's history, many of the owner-members are apt to say, "Well, seems like we had a couple of problems, but we worked them all out, and things are going real well now." The author heard that, or a variation of it, a hundred times if once. They are quick to downplay their own importance and to recognize the contributions of others. When someone else has done wrong, the cotton farmers are apt to attribute the failure to a lapse in judgment, or outside pressures, or something other than a character flaw. It seems there just aren't any "bad ol' boys."

The West Texans are among the peoples of the world who live in such powerful environmental conditions as to be shaped by them. Their humility is a result of the relationship they have with the land. Whatever pride they feel at their mastery over the land is tempered with humility at the power of the forces of nature on the Plains. The most successful farmers, well aware of the amount of work that has gone into their success, always harbor the feeling that their accomplishments have been at the pleasure of God and nature. Far from bragging about what they have done, they are more apt to feel thankful. No matter how well they work the land, they do far less to it than it does to them (and for them). They know, too, that if one day they all moved away, it would not be long before the Plains looked as they did before the first settlers encroached on them. That in itself is a humbling thought.

West Texas can be beautiful. There is something awe-inspiring about being in a place so flat and open there is absolutely nothing between you and your Maker. When a red-orange setting sun casts rays of gold on that land, through streaks of fine purple clouds against a darkening sky, the sight is moving. As you watch the sun swallowed by the flat plains, the horizon seeming farther than the sun itself, and the only sound intruding on you is the wind or the cry of a circling hawk, you will be very happy for the experience and put the memory away for a time and place when you will need it again. Which, in West Texas, can be very soon. The flatness and the wind, which always are there, and the dust, snow, rain and mud, numbing cold and searing heat, which take turns being there, have a numbing effect that tries to break your spirit and often makes you wish you were somewhere else.

West Texans have no delusions about their land being a potential tourist Mecca — a place renowned for charm, culture, and fashionable living. It just is not that way. It has charm — a rough, unspoiled type of beauty. It has culture — the folkways and urbanity of the frontier with a dash of contemporary style. It has easy living — not plush, but unhurried and comfortable. It is not a place where salon-dwellers would want to spend their lives, but it is a place every American interested in a fuller understanding of what their country is about would do well to see and experience and enjoy. And it is very enjoyable.

There is a story in the local folklore that, when God was creating West Texas, He made a mistake and let the earth harden before He could shape it into hills, valleys, and rivers. Looking at His error, dry, barren, and windswept, He despaired at having to destroy the area, wipe it clean and start over. "I know what I'll do," He said, "I'll just create some people who like it this way."

The people He created constitute the greatest asset of West Texas. They are no better or worse than Americans anywhere else. They just seem a little different, if only in the degree to which they are open, warm, friendly to strangers, and courteous. In their business relationships, they evidence a deeply-ingrained sense of responsibility that seems to have withstood three of the most pervasive and disturbing of modern America's characteristics — the "fast buck" mentality; the need for ironclad contracts developed and backed up by armies of lawyers; and the loose, "anything goes" attitude toward performance on agreements once they are made. Obviously, being out on the frontier isn't all bad. West Texans surely are not all choir boys (and girls). They have their share of scoundrels and scandals. But they do seem to take seriously their professional responsibilities to others. When you strike a deal with someone there and shake hands, you do not feel compelled to count your fingers.

This book is about cotton farmers, and they almost invariably are men. Except for an occasional mention, the book offers a picture of West Texas that is nearly devoid of women. In fact, the West Texas cotton culture is a patriarchy. But it would be inaccurate and unfair to omit mention of the role of women in that society. From the time the Plains were first settled, women have been the civilizing influence there. And their influence is enormous. They not only are responsible for much of the culture in the area, but, like women in many other places, they bear a tremendous amount of familial responsibility, raising the children while men are out farming, or tending to cooperative business, or riding miles of fence line, or breaking horses. Many of the women of West Texas possess the legendary Texas beauty. They also are intelligent, articulate, well-read, and fashionable. They are to be admired. Pioneer women in spirit, they are thoroughly modern in outlook and refinement. Nonetheless, the lines between the sexes are very tightly drawn in West Texas, and they have been drawn by the men. One gets the feeling that it may be a very long time, for example, before a woman serves in a position of responsibility in the management of a West Texas cooperative. And few of those positions require the holder to have so much as seen a farm.

The author undertook this assignment with little knowledge about West Texas and the cotton industry. The experience has been one of learning by "total immersion," and he hopes the result will show that at least some learning took place. He completes this undertaking with a fresh store of knowledge and deep appreciation for persons very different from himself who have, nevertheless, welcomed him into their offices and homes and become his friends. The completion of this book is not the end of his West Texas experience, but only the beginning.

Men of Vision,
Years of Change

Prelude to American Cotton Growers

SOFT CROP IN A HARD LAND

The Great Plains of Texas

There are many states of Texas. Texans themselves, it is said, divide their state into seven distinct regions, each with its own environment, culture, and history. The one most steeped in the hard, heroic legends of the Old West is the Great Plains. The Great Plains region of Texas is a vast, L-shaped province west of the Gulf Coastal Plain, extending to the eastern slopes of the Rocky Mountains. It includes the Texas Panhandle, from the Oklahoma border southward through Amarillo and Lubbock. Its southernmost reaches include the oil centers of Midland and Odessa.

The Great Plains are further divided into the High Plains and Rolling Plains. The High Plains region is a mesa that rises abruptly from an elevation of 2500 feet at the well-defined caprock escarpment, known simply as "the caprock." From the caprock, the region very gradually slopes upward to an elevation of 4600 feet at the border with New Mexico. Although the actual boundary is indistinct, the northern area of the high plains is called the Staked Plains, or Llano Estacado, itself divided into the North and South Plains. The Rolling Plains begin at the base of the Caprock and extend south and east. The topography of the High Plains is flat. The Rolling Plains, true to their name, are seemingly endless, gentle waves of land, an ocean in shades of brown and green.

In late autumn, winter, and early spring, cold winds roll down across the top of Texas, pushing rapidly through most of the state, sometimes reaching as far as the lower Rio Grande Valley. Texans call the cold waves "northers" — blue northers, wet northers, or dry northers. The chief characteristic of a norther that distinguishes it from what other people call a cold wave is the sudden, dramatic fall in temperature. The decline may be 20 or 30 degrees in two or three hours.

On the Great Plains of Texas, the changes in weather come fast and hit hard. With few obstacles in its way but man's puny works, the force of nature is like an onrushing train. It can be frightening, even in those few minutes when a sudden change is building. Perhaps it is especially unnerving then, when the impact and direction of the change still are unknown. Watching the drama in the skies unfold, a human being feels small and vulnerable. For the onlooker, there is the fear that when the "train" moves through, he may find himself on its tracks. There is a threat for every season. Heavy spring downpours turn country roads into wheel-sucking quagmires. Marble-sized hail in spring, or sheets of sleet and snow in winter, blanket the land in minutes. The wind blows. And blows. And blows. At various times of year, swirling dust, sand, or snow obliterates the landscape, erasing any sense of direction and leaving its mark everywhere. Flash floods and tornadoes are a fact of life, and sometimes death. West Texas is not country for the timid nor those whose existence depends on a steady diet of warm comforts served up with tame predictability.

The story of Texas is one of people who have traded comfort and predictability for freedom and opportunity. It was true of the explorers, early traders, the first settlers, stockmen and cowboys, merchants who followed them, and oil wildcatters who came later. And it was true of the farmers, including those who worked the land for Texas's richest agricultural treasure, cotton.

Cotton Comes to Texas

Even though the discomfort and uncertainty are far less today than when West Texas was America's frontier, the life of a cotton farmer is a difficult one. Those who grow that soft and gentle crop do so in conditions that are hard and unforgiving. Like farmers everywhere, they live at the mercy of nature and fully exposed to the uncertainties of the marketplace.

Cotton was the foundation upon which Texas built its economic wealth. Plants of the genus *Gossypium* were native to the area, and the first Spanish explorers and missionaries were quick to see their promise. By 1745, the missions around San Antonio reported the production of several thousand pounds of cotton annually.

Prelude to American Cotton Growers

When Moses Austin arrived in Texas in 1820 to obtain a colonization grant from the Mexican government, he planned to support himself and his settlers by raising sugar cane and cotton. In his formal presentation to Texas Governor Antonio de Martinez in December 1820, he explained that he represented 300 American families, all sympathetic to Spain, who wanted to settle permanently in Texas and cultivate cotton, sugar, and corn. Martinez was impressed. He informed Austin that he would forward his petition and recommend its approval to the commandant general in charge of Texas. Around the middle of May 1821, Austin got word from Texas. The commandant general had approved his petition. He would be granted 200,000 acres of his choice.

Moses Austin died in 1821, long before his ambition could be realized. His son, Stephen, followed his father's deathbed request and only legacy by continuing the settlement of Texas. And he built upon it, asking Governor Martinez for an immense grant estimated at 18,000 square miles, some 11 million acres, in southeastern Texas. The young Austin divided his grants so that each 4605-acre property consisted of two units: a plot of farming land, about 177 acres, and a tract of grazing land, about 4428 acres. The crops planted that year, 1823, were good and included not only corn but also cotton, which would soon become Texas's main cash crop.

Within five years of the commandant's approval of the Austin colony, one of its members, Jared Groce, had built Texas's first cotton gin, thus earning from later generations the title "Father of Texas Agriculture." As Texas's foremost cotton grower and one of its leading stock raisers, Groce became the food supplier to Austin's first colony. Groce enlarged his huge original property of 44,000 acres by buying up land from neighbors, and in 1825 he began processing his cotton harvest in his own gin.

The colonists Stephen Austin induced to southeastern Texas in the early 1820s quickly realized the territory was in every way ideal for cotton culture. The bottomlands along its rivers were incredibly rich and yielded enormous crops, running as high as 2000 pounds of cotton per acre. Moreover, the riverside cotton lands afforded easy access to cheap transportation. Flatboats and steamboats collected the cotton harvest in 500-pound bales and carried their cargoes downstream to new ports, chiefly Galveston, which sprang up along the Gulf of Mexico. Here the bales were transferred to seagoing vessels bound for cloth-manufacturing centers in the northeastern United States and in Europe.

Lesser growers than Jared Groce also did well. One man with a good-sized plantation was said to have realized the equivalent of $10,000 – in those days a fortune – on one cotton crop. Most trade was barter, for cash was still woefully short. An observer reported in 1832, "There are in these districts many good citizens, many good livers, men of property, who do not handle five dollars in a year." Those who happened to have cash could lend it at 25 percent per annum.

Cotton culture boomed in Texas in spite of floods, droughts, a financial panic in 1837, and a severe depression in the early 1840s. The cotton crop, estimated at 10,000 bales in 1834, jumped to 58,000 bales in 1849. In the 1850s, when the population trebled and the number of farmers doubled, Texas became one of the world's major cotton producers; its cotton production increased tenfold, reaching 431,000 bales in 1859. This phenomenal success increased the demand for, and therefore the value of, prime riverside land suitable for the cultivation of cotton. In one three-year period in the 1830s, cotton lands rose in price as much as 400 percent, and in the next decade, a large, well-developed plantation sold for as high as $8.50 an acre. The average plantation was relatively small and was worked intensively by four or five slaves, whose purchase price averaged $600 in 1846. Though cotton prices fluctuated, it ordinarily cost the owner of a plantation about five cents to raise one pound of cotton, which he sold for eight to 10 cents a pound. After all his expenses were paid, the planter ended up with a comfortable annual income of $3500 or more.

With an end to all immigration barriers for Americans after Texas achieved independence from Mexico, a rising tide of settlers could roll south to the Rio Grande, and north and west across the endless plains, homesteading where they chose. They would plant vigorous towns, develop their cotton industry into one of the world's largest, and lay the foundations of a cattle empire that would become greater still. And in 1845, the embattled nation would triumphantly join the United States as its biggest state, then grow prodigiously in strength during the 15 years thereafter. These feats and more the Texans accomplished with their feisty, risk-loving, me-first, Texas style. For liberated Texas was nothing if not a land of rampant individualists. An old settler called it a "free-fighting, stock-raising, money-hunting country," and a newcomer found it "full of enterprising and perservering people."

Whereas East Texas was producing nearly half a million bales of cotton in 1859, West Texas would be a hostile, forbidding place for another two decades. The white man's culture came late to the Great Plains of Texas.

The Llano — Uninhabitable Land

A large part of the Great Plains of Texas is the Llano Estacado, or Staked Plains. The name is attributed to the early Spanish explorers, but its meaning is uncertain. Some say the area was so named because the explorers had to drive stakes into the featureless terrain or leave other markers to keep from becoming lost. Others say the caprock on which the Llano rests looked to the explorers like a fortress or a "stockaded" territory.

A flat, isolated plateau, the Llano is 32,000 square miles of short-grass prairie. In length, it is 250 miles from the breaks of the Canadian River to the north to the dusty chaparral of the Permian Basin to the south; in width, 150 miles east and west. It is crowned to the north, east, and west by the caprock that separates the plateau from the surrounding lowlands. The Llano Estacado is a remnant from the apron of outwash that stretched eastward following the uplift of the Rocky Mountains 65 million years ago. Early American explorers labeled the land a portion of the Great American Desert and predicted it would remain forever uninhabited.

The arid Llano Estacado can be a forbidding place. The climate varies among extremes and includes frequent, recurring cycles of drought. Relentless spring winds abrade the soil and, when the climate turns dry, they extend to summer. Sudden, severe storms in spring and summer bring torrential downpours, hail, and frequent tornadoes. But the land contains extremes of beauty as well: incredible sunsets; an exhilarating summer climate; a bountiful resource base in the soil.

Herds of pronghorn and huge herds of bison made the immense grasslands of the Llano an attractive hunting ground for nomadic Indians who utilized the game for subsistence. Later, those nomads took horses from Spanish settlements in the Southwest and enjoyed an incredible freedom on the open plains. With the horse, they hunted bison with greater ease and gained hegemony over the Llano. But to white settlers, the Llano was inhospitable. Then, as now, it lacked shelter from natural and man-made dangers. There was room to run on the Llano, but no place to hide. The region was home to members of the proud, unyielding Comanche tribe of Indians. The Comanches were reputed to be the best horsemen on the American plains, fighters of unsurpassed courage, their warrior bands a wind of fury. The Comanche ruled the Staked Plains.

The legendary Comanche chief, Quanah Parker, was one of the last holdouts against white encroachment. He and his braves made days watchful and nights sleepless for settlers who ventured into their homeland. Quanah was a half-blood, born of a Comanche father and a white mother, Cynthia Parker, who had been kidnapped by Comanches when she was nine. He was a member of the Noconi (Those Who Turn Back) band of the Comanches until the late 1850s, when his mother was recaptured by whites and his father and brother died of illness. With no reason to stay in the Noconi, Quanah joined the powerful Kwahadi (Antelope), a band that lived on the edge of the Staked Plains. He probably was attracted to the Kwahadi by their reputation as persistent and skillful raiders. If so, he made the switch at an opportune time, for the Comanches suddenly found themselves almost unopposed. The Civil War not only stripped the forts of U.S. soldiers, but also sent about 60,000 Texans into the Confederate Army, leaving scarcely 27,000 men behind to defend the entire state.

The Comanches, with the Kiowas and other allies, turned Central Texas into a killing ground. Hundreds of settlers were massacred and their settlements reduced to charred ruins. When the U.S. Army returned after the war, an officer remarked, "This rich and beautiful section does not contain today as many white people as it did when I visited it 18 years ago." Quanah and his braves did not succumb until 1874, and then only under relentless pressure from the U.S. Army and a flood of settlers. The white man subdued the Comanche by eliminating the bison and the war pony. White buffalo hunters, using horses, wagons, and powerful rifles, exterminated the bison, the Indian's main source of sustenance. At the final battle of the Indian Wars in Texas, U.S. Cavalry soldiers surprised a large Indian encampment in Palo Duro Canyon near what is now Amarillo. They killed only five braves, but they captured 1400 horses, selling 400 and killing the rest. For 10,000 years, the Llano Estacado had been a natural game preserve supporting nomadic hunters and gatherers. That era passed within a decade. With neither food nor freedom, the Indian had no choice but to lay down his arms.

So menacing was the Llano that it found a place in Larry McMurtry's epic novel, *Lonesome Dove*. He painted a terrifying picture of a land of sullen desperadoes and sudden death. Of his book's main antagonist, the hateful renegade Blue Duck, McMurtry said, "[He] stole white children and gave them to the Comanches for presents. He took scalps, abused women, cut up men. What he didn't steal he burned, always fleeing west onto the waterless reaches of the Llano Estacado, to unscouted country where neither Rangers nor soldiers were eager to follow."

In time, this land thought in the 1850s to be uninhabitable was tamed. Entrepreneurs turned the prairie into a stockman's paradise until prices softened and ranchers made money selling settlers land instead of meat. Those settlers were self-sufficient stock farmers who raised feed to supplement the range upon which their cattle sought sustenance. The plow followed the stock farmer, brought west from the Blackland Prairie or Western Cross Timbers by emigrants who fled the ravages of the boll weevil and sought inexpensive lands and a chance on a future out West.

Between 1910 and 1930, many large ranches were subdivided and sold through land-colonization schemes. Promoters used national advertising, excursion fares, testimonials, and demonstration plots to entice prospective buyers to the Llano Estacado. The land promotion efforts attracted an experienced, agrarian class and convinced the new settler that the plains were, with work, a potential agricultural paradise. The Llano evolved from a region of large ranches and scattered pockets of subsistent settlement to small, cultivated parcels of family farms. The uncertainties of life on the open plains encouraged close cooperation among neighbors. Part of this cooperation was founded on the arid lands' code of hospitality. Part of it was the cohesiveness that arises when individuals work together to build social institutions in a new setting. Somehow, Texans found a way to maintain their style of "rampant individualism" while working together toward a common good.

The Llano Estacado stimulated this strong sense of community among new residents. Anglo society evolved along agrarian lines with an emphasis on education, family, religion, and community service. This spirit of community service later became one of the underpinnings of the cooperative agricultural movement. It was the spirit that moved men to answer the call of their neighbors, to serve on boards and committees of cooperative enterprises, donating substantial amounts of their time and effort with no remuneration other than their communities' admiration and respect. Social life was an appealing distraction from the hard work of farming. Neighbors organized community singings, picnics and other activities that reinforced bonds of cooperation on the open plains. As villages grew into towns, residents organized women's groups to disseminate practical information among neighbors and to provide intellectual stimulation through study clubs. These groups later provided leadership for projects that improved the quality of life in frontier communities.

Cotton on the Plains

The new émigrés "broke out" prairie in 12-inch strips and developed an intimacy between soul and soil as they walked the freshly turned sod in the company of stubborn mules. It was inevitable they would begin the cultivation of cotton, the cultural practice that sustained their forebears in the lands "Back East." Many of the cotton producers of today's West Texas are the children of two migrations westward. The first was from the southeastern states to East Texas in the period from before the Civil War to the early part of the twentieth century. The second was from East Texas to West Texas during and after the early part of the twentieth century.

In 1903, Floyd and Hale counties sported new gins to separate the cotton from its seed and put it in bales. Lubbock County followed a year later. By 1915, Crosby, Dawson, Floyd, Lubbock, and Lynn counties produced 14,839 bales of cotton. Remarkably, their production was more than 800,000 bales in 1949. Cotton became the "money" crop on the South Plains. Yet cotton on the Plains was not the cotton of the South. In the South, cotton culture had been a feudal industry, labor-intensive and mistrustful of change. In contrast, the Plains was an agricultural democracy. Everyone was equal. No one knew how to farm the semiarid Llano Estacado when he arrived; drought and economic dislocation in the 1930s underscored this truth tragically.

Nonetheless, the new class of farmer was young and progressive. Cheap lands drew him to the Plains where he found freedom from the restraints of the old farming districts. There, things were done a certain way, if only because it had been the way of his father, his father's father, their grandfathers, and so on back through the corridors of time. Through a troublesome combination of necessity and inclination, the farmer who evolved on the Plains developed a healthy propensity for innovation. He learned to accept change, or he failed. He extolled the doctrine of unending progress, and as a matter of practicality, kept one eye on his neighbor's crop and cultivation practices and adopted those that proved financially promising. He was no stranger to science.

When the Depression revealed the impropriety of poor cultivation practices, he heeded the advice of agronomists and diversified. By the late 1930s, cream and egg sales generated cash flow and groceries for rural families, while Lubbock, Abernathy, and Plainview became dairy centers.

Economic success, however, came after World War II when diversification was abandoned, and the farmer planted cotton, cotton, and more cotton. Here, where the horizon levels the land, the small family farmer found prosperity at last. Or so he thought. The crop lost its economic luster in the mid-1950s, and a tightly bound consortium of farmers, agronomists, and textile engineers struggled to keep cotton farming alive. By then, commercial agriculture had supplanted the self-sufficient family farm.

The transition to a cotton economy occurred for several reasons. Prices were good after the war. Irrigation guaranteed a crop. Mechanization enabled a single producer to cultivate 450 acres or more of cotton. The cotton district on the South Plains grew to cover a 13-county region bounded by Bailey, Gaines, and Floyd counties. In 1947, as the Texas cotton belt shifted west, the South Plains alone marketed 900,000 bales of lint cotton, 28 percent of the state's total. By 1977, 77 percent of the Texas cotton crop originated in the Texas High Plains and Rolling Plains, with the majority of production in the counties surrounding the city of Lubbock. The 4.2 million bales produced that year represented 30 percent of the total U.S. crop.

"The Cottonest City in All the World"

Lubbock, seat of Lubbock County, is the industrial, financial, and commercial hub of the south plains area of West Texas. It sits about 40 miles west of the edge of the caprock, about 100 miles south of Amarillo. The first settlers were ranchers who arrived in the late 1870s. The town was laid out in 1891, when Lubbock County was created out of Bexar territory and named for Tom S. Lubbock, a Texas Ranger and Confederate soldier, who was the brother of Francis Lubbock, Governor of Texas during the early part of the Civil War. Incorporated in 1909, the town adopted a mayor-commission-manager form of government in 1917.

Lubbock became the capital of the South Plains cotton district. The city had an ambiguous introduction to boom times in 1947. Things looked bad that summer. Rainfall quit in June, and dryland farmers feared another crop failure. A shortfall in the American cotton crop that autumn sent American textile mills scrambling for whatever cotton the South Plains produced. Traditionally, West Texas cotton was exported to Germany, Italy, or Japan where spinning techniques were better suited to handle the shorter, finer fiber. Domestic demand brought instant prosperity for what proved to be a bumper crop.

As the crop was moved to storage it swamped warehouse facilities. The Texas Compress and Warehouse Company of Lubbock and affiliates, for example, had a storage capacity in excess of 500,000 bales. By late fall, the warehouses were filled, and workers began storing bales on the ground in the yards adjoining the facilities. Warehouses in Lubbock prorated the amount of cotton they accepted from gins on a daily basis. In some cases, warehouses refused additional shipments for storage until the yards and warehouses were cleared. In 1947, most cotton was shipped by rail. Railroad officials circumvented a shortage of rolling stock by loading cotton bales in every conceivable vehicle, including refrigerated cars and gondolas to get it to market for the textile mills of the Deep South. When the gin dust cleared, Lubbock County, with a farm cash income of $32.8 million, ranked third among Texas counties in agricultural receipts for the 1947 season. Farmers in counties on the Plains received a quarter-billion dollars in cash income and led all other sections of the state.

Thirteen years later, the Lubbock Chamber of Commerce convinced the editor of the *Cotton Gin and Oil Mill Press* that Lubbock was the "Cottonest City in All the World." Lubbock officials in 1960 could point to a complex of compresses, oil mills, and gins representing a $113 million investment. The cotton complex directly employed 2885 people year round although, in peak production, jobs approached 11,000 with a payroll of $44.4 million. Furthermore, Lubbock was a retail center for the estimated 130,000 migrants who earned $45 million harvesting cotton each fall.

Lubbock also was home for research institutions that followed the cotton plant from breeding through production, ginning, and the manufacture of textiles. Traditionally, the Llano Estacado rewarded creative innovation. In the twentieth century, those rewards occurred in the marketplace. Farmers demonstrated a tendency for innovation on the Llano Estacado and used technology such as mechanization and irrigation to increase greatly the productivity of the region for commercial gain after World War II. Plant breeders working at experimental stations, for private industry, and on local farms altered the nature of traditional crops to fit an energy-intensive, highly mechanized form of scientific agriculture. Meanwhile, Llano Estacado farmer associations created tools such as instrument cotton classing and electronic marketing to open new domestic and international markets for the region's productivity.

Lest anyone underestimate the pace of change on the Texas Plains, consider how dramatically the cotton industry, and with it the way of life, accelerated in the little more than 100 years after the Comanches gave up their hold on the region. In less than half that time, from 1935 to 1980, improvements in energy distribution, transportation, technology, and the economy elevated the producer's standard of living exponentially. It was only a few decades ago that the producer's parents arrived via wagon to the Plains where they built a box-and-strip house and spent long days behind a mule. Suddenly, or so it seemed, these same people found themselves in homes with electricity and indoor plumbing, while their children rode upon machines that cultivated four, six, or more rows of cash-crop cotton in a single sweep across the field.

The story of L. C. Unfred, chairman of American Cotton Growers, is typical. It spans the old West Texas and the new in a single lifetime. Unfred was born in 1924 in Runnels County, Texas. He was two years old when his family moved to the High Plains. His father had been a farmer in central Texas, then in Runnels County, when relatives told him, "The High Plains is the country; it's the place to come."

"You've heard the old story," Unfred says, "moving west." Setting out in a T-Model Ford pulling a trailer with all their possessions in it, they reached the base of the steep caprock at Post, Texas on Thanksgiving Day, 1926. "If you talk to the old-timers," Unfred says, "Thanksgiving Day, 1926, was the worst sandstorm on record for this entire area." Anyone familiar with West Texas weather will know immediately how bad it must have been. "The wind was howling and the sand was blowing and the old T-Model wouldn't pull the trailer up the caprock. [The car] had more power in reverse," he says, "so my dad unhooked the trailer and tied it to the front bumper of the old T-Model. Mama drove it . . . she just 'mashed' the reverse pedal and backed it up the caprock. My dad and brother were out pushing on the trailer to get it up the hill.

"We got up to the top of the caprock, and Mama said, 'Look, it's snowing!' The ground was just white. We got to looking and it was cotton, cotton that had blown out of the fields during the sandstorm.

"The sand was so bad, it was dark in the day," he says, "but finally we got to where we were trying to go for the night. We stayed with these neighbors for a night or two. Then we lived in a tent from a few days after Thanksgiving until probably March or April of the next year. We spent the winter in a tent, the four of us, until we built a little, two-room 'shotgun' house and moved into it in the spring."

Unfred vividly remembers how cotton was harvested before mechanization. "We used to pull cotton by hand. We'd 'turn out' school for two or three weeks in the fall, and all the kids and everybody would go out and harvest. You'd pull each boll off by hand," he says, "and put it in a sack that you had over your shoulder. You were dragging this sack and pulling cotton, then take it, weigh it, and dump it in the wagon. When we had enough to make a bale, 2000 pounds or so, we'd put it in a wagon that carried two bales to the gin. It would take probably a week or longer of us four pulling to get a bale to take to the gin." It was one of those experiences that can be enjoyed only as a memory. Unfred remembers, too, when electricity first came to the family home. It wasn't until he was 16 years old, less than 50 years ago.

AGENTS OF CHANGE

George Mahon — Quiet Power in the Congress

Change on the Texas Plains was rapid but not random. It was resolutely and carefully engineered by men and women of vision. One of the most significant of these was George Herman Mahon. Mahon went from rural Texas to Washington, DC, as a young man to represent his fellow West Texans in the U.S. House of Representatives. He stayed for nearly 50 years. In his time in the Congress, Mahon shaped not only an impressive career in public service but laws, policies, and programs that will forever affect the district he served. While West Texas cotton farmers were working at home to build an industry, Mahon was backing them up in the halls of Congress.

He was born to John Kirkpatrick and Lola Brown Mahon on September 22, 1900, in Mahon, Louisiana. He moved with his family to Mitchell County, Texas, in 1908 where he grew up with his seven brothers and sisters on a farm near the community of Loraine. A graduate of Loraine High School, Mahon continued his education at Simmons College in Abilene and the University of Texas Law School in Austin. In 1925 he and Helen Stevenson Mahon, his wife of two years, settled in Colorado City, Texas. There he practiced law. Mahon's career in public life began in 1926 when he was elected county attorney of Mitchell County. In 1927, Governor Dan Moody appointed him the Thirty-second Judicial District Attorney, a position he held for three two-year terms. In 1934, Mahon campaigned successfully against seven opponents to become the first Representative of the newly created Nineteenth Congressional District. It was the last job he ever had. He served continuously in that office until his retirement in 1978, having served longer at that time than any other sitting member of Congress. George Mahon understood Washington and the way it works. He went about the business of public service in a quiet, unassuming way, playing by the carefully crafted, long-established rules of the Congress and amassing enormous power and influence along the way.

A story from Robert A. Caro's book *The Years of Lyndon Johnson: The Path to Power*, is illustrative. Caro contrasts George Mahon's style with that of another Texan, one considerably more flamboyant and not nearly so successful until many years later, when he not only learned the rules but was in a position to use them. The incident recalls Lyndon Johnson's efforts to break out of the "seniority trap" in 1938, just one year after he came to Congress.

> The only House committee in which a junior member was anything more than a cipher was Appropriations, and of all the House committees, only Appropriations had the power to fund government programs. . . .
>
> In 1938, the traditional "Texas seat" on Appropriations fell vacant, and although George Mahon wanted it and had seniority, Johnson tried to step into it, planting newspaper stories hinting that his "close [Roosevelt] administration contacts" would enable him to use the Appropriations post to obtain more federal projects for Texas.
>
> But he never had a chance. The Texas seat would be filled by the Texas delegation, and on that delegation, Mahon [recalled], "Mr. Rayburn had the power." Although Mahon knew of Rayburn's paternal fondness for Johnson, he never worried about the result.
>
> "Whatever Rayburn said went," [Mahon said]. But Rayburn followed the rules. And the rule that mattered was seniority. "I was the senior Texan who wanted the spot. I was in line for it. If you were the next man in line, you got it – that was the way the unvarying rule was."
>
> Mahon got it – found the immediate rewards he had expected ("Even as a new member of Appropriations," he [said], "if you were on Appropriations, you were courted by other members"), and he began the

climb to greater rewards: in eleven years, he would be chairman of his subcommittee, in twenty-six years, at the age of sixty-four, chairman of the full committee; for his remaining fifteen years in Congress (until he retired at seventy-nine), he would be a power on Capitol Hill. Johnson didn't get it; he remained a member of the Naval Affairs Committee– a junior member of a committee on which junior members were limited even in the number of questions they could ask.

Mahon chaired the powerful 55-member House Appropriations Committee from 1964 to 1978, where he advocated a strong national-defense program and stressed the need for fiscal conservatism. Mindful of the agricultural base of his Nineteenth District, he played a key role in the development of farm programs and energy measures throughout his years in Congress. One of his major achievements on behalf of his constituents was cosponsorship of the Rural Development Act of 1972. The author of the measure was a Texas colleague and personal friend of Mahon's, Congressman Bob Pogue, whose name is legend in agricultural America. From Pogue's seat as chairman of the House Agriculture Committee, and from Mahon's as chairman of the Appropriations Committee, Texas agriculture was well served. The business and industrial development section of the Rural Development Act empowered the U.S. Department of Agriculture, through its Farmers Home Administration (FmHA), to provide loan guarantees for projects to stimulate jobs and economic enrichment in rural America. By encouraging and supporting the development of commerce and industry in rural areas, it recognized the limited ability of agriculture to alone sustain prosperity and growth in those areas. By providing options and opportunities for rural Americans and keeping small communities alive, it answered the question, "How ya gonna keep 'em down on the farm?"

Roy Davis – Visionary Father

Mahon's careful cultivation of the nascent agriculture industry in his district brought him into close and frequent contact with agricultural leaders. Probably none of these leaders engendered more respect from, or made a stronger impression on, George Mahon than Roy B. Davis. In Roy Davis, Mahon had a strong supporter. In George Mahon, Davis had an ally in the halls of political power. In each, the other found a friend. It probably is not coincidental that the largest project to receive federal support under the Rural Development Act of 1972 was a textile mill in Mahon's district. The mill was a life-long dream of Roy Davis. In

The First Forty Years, A History of the National Cotton Council of America, 1939–1979, Roy B. Davis, the seventeenth president of the Council, is described. "If the conversation was about cotton and the name 'Mr. Roy' was mentioned, no further identification was needed. If ever any one man epitomized the best in one great area of the country, it was Roy B. Davis of Lubbock and the High Plains of Texas."

Davis was born on December 10, 1900 (just 79 days after his contemporary, George Mahon), in McGregor, Texas. He was one of 10 children, four boys and six girls. As a teenager he was a 4-H Club member in Dawson County, about 60 miles south of Lubbock, with "a few acres of cotton and corn."

His son, Dan, recalls that one of Roy Davis's early jobs was at the textile mill constructed by C. W. Post, the "Cereal King," at Post, Texas. That experience may have influenced some of his thinking in later life. When he left that job, he also "left home and came to Lubbock, lived with a relative, and went to high school." According to the younger Davis, his father graduated from high school "at a relatively late year in life and then worked his way through Texas A&M." Roy Davis earned a degree in dairy husbandry there in 1927. He married Dennise Cobb on December 21, 1929, a marriage that would last nearly 46 years, the rest of his life.

Before becoming general manager of the Plains Cooperative Oil Mill in 1943, where he later achieved legendary stature in Texas agriculture, Roy Davis was a county agricultural agent in Terry County and Gaines County, and then went to Plainview as county agent of Hale County. Plainview is where Dan Davis was born in 1930. His father managed a cooperative creamery there, "very successfully back in the Depression.

"In '36, I guess it was, our family moved to Houston where Dad became secretary of the [Houston] Bank [for Cooperatives] from '36 to '43. But my mother and father didn't like Houston. They preferred to be [back in Lubbock], so when the oil mill was looking for a general manager, Dad came back . . . in '43 and became the general manager." Under Roy Davis's guidance, Plains Cooperative Oil Mill became the largest cottonseed processing plant in the world. Even more important, from his position there he envisioned, inspired, and supported an array of programs that increasingly gave the cotton farmer more control over his livelihood. After retiring from the oil mill in 1973, Davis became the interim director of the Textile Research Center (TRC) at

Texas Tech University in Lubbock, better known as Texas Tech. It was not without justification that "Mr. Roy," the name by which he was known and greeted by citizens of the Lubbock area, also was known as "Mr. Coop" or "Mr. Cotton" by his industry.

Wayne Martin, now manager of the Plains Cooperative Oil Mill, the same position from which Davis accomplished so much on behalf of the cotton farmer and from which he spawned so many innovative ventures, recalls the man. "He was very dedicated, hard-working, almost with a one-track mind as far as doing something to help the producers in forming cooperatives. He really was interested in seeing that the cotton producer moved his products closer to market, and he worked on that from day one, from the time he went to work [at the oil mill] until the day he died. He was a very likable fellow," Martin adds. "He could relate to the farmers out here and to members of Congress. He could talk to the farmer on the turnrow or talk to the President of the United States. He treated them both pretty much the same."

Jim Parker is director of the Textile Research Center at Texas Tech. A protégé of Davis, he remembers with some awe that the older man's associations extended well beyond that with George Mahon. "John Connally was one of his close friends. And he and Lyndon Johnson were fairly close." Davis's powerful friends were not the only reason for Parker's admiration. Davis was a mentor to him, "one of the best friends I've ever had." He talked the young Parker into returning to West Texas and then literally brought him home with him.

Parker, a North Carolinian, had lived in Lubbock in the early 1960s. He went there after several years in the textile industry in his native Southeast, followed by four years in Latin America, building and running a textile school on a U.S. State Department project. Of four job offers Parker received when he returned from Latin America, he and his wife chose the one from Lubbock. They did not want to return to the east coast. According to Parker, in an interesting commentary on West Texas, "we found that life in Lubbock was more nearly like the life we had led in South America." They returned to North Carolina a few years later. But by 1973, "Mr. Roy" talked Parker into coming back to West Texas and running the Textile Research Center. "He just insisted," says Parker. "He was a hard-headed old fellow. I loved him. But he was bull-headed and you couldn't change his mind at all."

Parker provides one view into how Davis became interim director of the TRC after his retirement from the oil mill. "He had been manager of that oil mill for probably 30 years. When he retired, they hired a new fellow to run it. But he wouldn't leave. He'd stay there every day to try to tell the guy how to run it. The guy said, 'He's driving me crazy,' so they gave him an honorary position at Texas Tech and gave him an office out there just to get him out of the oil mill." For the next few years, Parker worked side by side with Davis. "Every afternoon he was wanting me to go home with him. I was out here by myself at that time because my family was still in Charlotte. He'd take me home and we'd sit down and chat." It was through this close, personal contact that Jim Parker developed much of the love and esteem in which he holds Davis to this day.

Whatever the circumstances of Roy Davis's arrival at the TRC, there is no question that the university was glad to have him and the knowledge, experience and prestige he brought with him. He had a deep and genuine interest in research. In fact, his service at the TRC would crystallize his thinking about textile production and help propel the development of that industry in West Texas.

At one time or another, Davis was just about everyone's "Man of the Year." *Progressive Farmer* magazine and the Texas County Agricultural Agents' Association so honored him. The Federal Land Bank presented him with its Golden Anniversary Medallion; Future Farmers of America and the Texas Farm Bureau Federation presented him with their Distinguished Service Awards. There were many others to come his way, including the presidency of the National Cottonseed Products Association and service as secretary of the Cotton Board, a national industry association. In a tribute paid Roy Davis late in life, his old friend George Mahon gave him yet another title. Mahon dubbed him "Mr. Humility," because he never had sought personal glory or credit. He concentrated instead, in Mahon's words, "on just getting the job done."

Conservative and persevering, Davis might be perceived as hidebound. That would be a mistake. Rather than egocentric, he was self-effacing. Rather than a misanthrope, he was a man of the grass roots, a consensus-builder who had the rare ability to pull people of divergent views together, and whose enthusiasm was as contagious as it was sincere. Rather than myopic in his interests, he was a visionary who saw the conditions, demands, and possibilities of the future and worked to build the technical and political means for meeting them head-on.

Examples abound. One of his concerns was expansion of cotton exports. In the National Cotton Council, he chaired a special export committee that analyzed U.S. potential and mapped strategies for the industry to follow. His personal dedication to long-term improvements in exports of cotton and cottonseed products served the industry well for many years. He also was at the forefront of those wanting to strengthen cotton's position with the domestic mills. Man-made fibers had been gaining in popularity, and cotton had to meet them in the marketplace on equal or better terms. It meant constantly improved quality, more promotional clout, adequate supply, and a competitive price. Roy Davis worked hard to achieve those ends. Accorded one of the highest honors of his industry, the presidency of the National Cotton Council, in 1968, he became board chairman in 1969 and then advisor to the board from that time until his death in August, 1975. No single figure was more important to the development of cooperative agriculture as a force in West Texas.

WHY COOPERATIVES?

For the Common Defense

"If you want to know 'Why cooperatives?' look at Lubbock," said the August, 1965, issue of "News for Farmer Cooperatives." The history of West Texas cotton in the postwar era is largely the story of how the region recognized its problems and organized to solve them. At the center of the story is the success of two organizations headquartered in Lubbock. Plains Cotton Growers, Inc. (PCG) and Plains Cotton Cooperative Association (PCCA) worked to develop a niche for South Plains cotton in domestic and export markets. Saving the cotton industry on the Plains was the "great cause" of Plains Cotton Growers, Inc., which organized in Lubbock in January, 1956.

Local producers were up in arms over state and national attacks on West Texas cotton. West Texas cotton is short-staple cotton, and was generally considered inferior to longer staple cotton. (Staple refers to fiber length.) Much of that reputation was fostered by competitors in other parts of the country and even within Texas itself. The state Agricultural Stabilization Committee brazenly had lifted allotments for crop subsidies from the Plains and South Texas, redistributing the acreage in East and Central Texas. Meanwhile, Congressional representatives from other cotton producing regions were proposing legislation to change the official definition of average cotton in the government programs to fiber one inch in length, longer than average West Texas fiber. The first item took cotton acreage from the Plains; the second would have financially penalized the cotton raised there.

PCG included 23 counties in a region bounded by Deaf Smith, Howard, and Motley counties. Its stated goals were service, research, and promotion. In its first newsletter to producers, PCG vowed "to provide accurate information to legislators; to finance research to improve the quality and quantity of cotton while reducing costs of production and marketing; and to combat the discrimination which existed against the cotton raised on the Plains." The prejudice stemmed from the cold statistics surrounding West Texas cotton. When PCG organized, the average staple length for South Plains cotton was just over 15/16ths of an inch, with a strength measurement that was less than ideal. In the textile mills, West Texas cotton presented problems that caused production delays and cost time and labor in a highly competitive industry.

Mills in the 1950s used a ring-spinning technique that demanded longer staple cotton than was grown in West Texas. The short staple was one reason West Texas cotton went for export, into government warehouses, or sold for ridiculous discounts. During the surpluses of the mid-1960s, for example, South Plains cotton accounted for 61 percent of the five million bales of short-staple cotton in government storage.

Roy Forkner, a progressive gin owner in the Canyon community, was active in PCG. Upon becoming president, he was sent on a fact-finding mission to interview mill owners and learn why the Plains received the lowest price in America for its cotton. The textile people with whom he met expressed dismay over the wide variation in the staple length of the cotton produced on the Plains.

Dallas Brewer, an American Cotton Growers board member, talks about the difference between West Texas cotton and cotton grown further west. "In California, they raise what they call the 'picker-type' cotton, an Acala variety that will grow five feet tall. On the other hand, [our cotton is a 'stripper-type' cotton, and if it] gets three feet tall, it is too tall. A short stalk with a lot of bolls is what we go after; lots of cotton on it. We have a lot of wind here," he adds, "and in the fall of the year, when that cotton opens, if we raised the type cotton they raise in California and Arizona, it would all be on the ground before you could get it harvested. Ours is a storm-proof cotton. It's shorter, the burrs are tighter, and it will hold the lint in."

The mill buyers knew the cotton by knowing the region from which it originated. With Plains cotton, the story was different. The farmer grew short, stripper cotton; however, he grew as many as three varieties of it in his field. He harvested all of it together, and the result was a wide variation on a bale-by-bale basis across the whole region. South Plains cotton was a headache for the mill owner. Only low price made it competitive.

PCG set out on local and national levels to alter the realities of West Texas cotton. In 1958, it cooperated with the Texas Agricultural Extension Service and the Texas A&M University Experiment Station on a quality-education program for Plains farmers. More than 2100 radio programs on 12 stations and articles and advertisements in 50 area newspapers carried the message of longer, stronger fiber and better handling during harvest. The programs broadcast advice weekly during the cotton season, covering every topic from planting and cultivation to harvest and ginning.

Further PCG efforts returned some of the allotment acreage lost to East Texas in 1956 and provided $55,000 to the United States Department of Agriculture (USDA) to humidify the cotton classing offices in Lubbock, Brownfield, and Lamesa. Humidifying these offices made the results of classing more accurate, helping ensure that West Texas cotton would not suffer from a faulty classing process in addition to its own shortcomings.

Equally important, PCG provided leadership in two programs that helped sustain the cotton industry on the Plains. First, it launched a counterattack against the boll weevil, which had begun to threaten area cotton crops. PCG coordinated meetings with entomologists to identify problem areas, developed control measures, and arranged financing for a successful decade-long attack that began in 1964 on the cotton pest. Second, PCG participated in efforts to finance and reactivate the pilot spinning plant at Texas Tech in 1958.

The pilot spinning plant conducted tests that benefited producers, breeders, and mill operators and examined in detail the effects of fertilizer, irrigation, and ginning processes on the end-product qualities of various cottons. Spinning techniques pioneered at the pilot plant found ways to reduce waste, utilize low micronaire (fine) cotton, and provide technical advice on recommended processing speeds, settings, and yarn size to be used when spinning Texas cotton.

The pilot spinning plant at Texas Tech became the Textile Research Center, which by 1970, could look back on a long and fruitful relationship with Plains Cotton Growers. PCG had financed 4000 square feet of floor space, paid for the equipment that maintained consistent temperature and humidity, and had even furnished the draperies on the walls of one of the most attractive office complexes on the Texas Tech campus. In return, the TRC provided accurate data that opened markets for a cotton fiber that was once the last choice for mill owners. The TRC also was involved in the development of automated fiber testing in conjunction with Plains Cotton Cooperative Association (PCCA) and Motion Control, a Dallas electronics firm. The push for instrument evaluation of cotton fiber centers to a large degree about PCCA, a farmer-owned marketing agency headquartered in Lubbock with worldwide outlets. PCCA is an outgrowth of another cooperative association in Lubbock, the Plains Cooperative Oil Mill (PCOM).

The Lubbock Cooperative Family

Like Plains Cotton Growers, both PCOM and PCCA represent an agrarian tendency, markedly pronounced in West Texas, for farmers to band together into associations for mutual benefit. Unlike Plains Cotton Growers, however, the oil mill and PCCA are cooperatives. That is, they are associations providing services at cost for the members who patronize them. In practice, such services usually translate into additional money for the producer and enable producers to control additional aspects of their livelihood.

The progenitor of the large cooperatives in Lubbock was Northwest Cotton Growers Association, which organized without capital in the mid-1930s to market South Plains cotton. Following its dissolution, a second organization, Plains Cooperative Gins, Inc., was chartered in 1936 to sell cotton for gins and farmers. In 1937, the cooperative bought an oil mill plant in Frio County, transported the machinery to Lubbock, and with a loan from the Houston Bank for Cooperatives, began operations in October, 1937. That facility became the Plains Cooperative Oil Mill.

Plains Cooperative Oil Mill. For those unfamiliar with cotton, the term "oil mill" might require explanation. It is a place where cottonseed is taken after being separated from the cotton (called the lint) at the gin. The oil mill breaks down the cottonseed into by-products and ships them to buyers. The cotton industry is not one to waste materials or overlook their profit potential; the margins in cotton are not large enough to permit it. Of every 2000 pounds of cotton that are pulled off the stalk, about 500 pounds are white lint, the pure cotton that leaves the ginning process and is turned into fabric. Another 600 to 800 pounds are burrs, sticks, stems, dirt, and dust, which are separated from the lint in the ginning process. The remaining 700 to 900 pounds are cottonseed.

Four by-products come from cottonseed. The first is cottonseed oil, which has numerous uses in food and other products. The second are the "linters," tiny pieces of cotton fiber that have stuck to the seeds after they have been separated from the lint at the gin. Linters are used in "chemical lint" that goes into plastics, paper, and printing ink. The third are the "hulls," the shells of the cottonseed. They are shipped to feedlots to be mixed into animal feed. They sometimes are used in synthetic materials, too. Finally, there are "cake" and "meal" made from the meat of the seed. These are used in animal feed. Little of the cotton plant goes to waste.

The Plains Cooperative Oil Mill was organized in 1937 because the price of cottonseed in the Lubbock area had historically been much lower than in other sections of the state. Only 22,000 tons of seed were processed in the first two years of the cooperative, but the venture returned dividends to its member gins. The early years of World War II brought losses and poor patronage until, in 1943, the oil mill hired Roy Davis as manager.

Wayne Martin recalls that "Roy Davis came to this mill from the [Houston] Bank for Cooperatives. He really got out and began to get some support for the mill and probably the early years were spent as much as anything in going into communities in this area and helping the farmers to form cooperative gins."

Dan Martin, director of field services for PCCA and a cousin of Wayne Martin, explains, "The Plains Cooperative Oil Mill struggled along for two or three years and almost went under. [The farmer-owners of the oil mill] hired Roy to run it, and he did an outstanding job with it." According to Dan Martin, the industry tried to "beat down West Texas cottonseed" just as it had West Texas cotton. The area's detractors claimed its "cottonseed was inferior to cottonseed raised in other areas of the state of Texas and in [other parts of the] cotton belt of the United States. Through the Plains Cooperative Oil Mill, Roy proved that the products coming from that seed were not inferior, in fact they were superior. The independent segment of the industry was using the 'inferior' [claim] to beat [PCOM] down on price. "That oil mill," he continues, "in essence was organized and built in self-defense, trying to protect the farmers' own interests. Roy Davis made quite a name for himself through that work."

Davis, a tireless, relentless worker, increased the participation of gins in the oil mill and oversaw improvements in the way the mill operated. In 1946, thanks to the foresight of Davis and the pressures of wartime shortages on prices, the mill went from a deficit of $2394 in working capital to a surplus of $143,467. Machinery advances in the early 1950s tripled tonnage to 102,376 by 1953, and the mill built a substantial capital surplus by the early 1970s.

Lubbock became an oil-processing center as early as the mid-1950s. The city had three mills that turned cottonseed into oil, animal feed, and related products and anchored the industrial corridor along Avenue A on the city's east side. PCOM was, by far, the largest. In fact, it was the largest such oil mill in the world.

PCOM, for so long a model of successful management, nearly went under after Roy Davis retired. In the winter of 1974, a short cotton supply and low cotton prices had muddied the future of cotton on the Plains. Meanwhile, the sunflower-seed business was bullish worldwide. In January, 1975 the oil mill contracted for 223,000 acres of sunflower seeds, promising South Plains farmers 15 cents per pound for the product. Unfortunately, the mill failed to secure a market for sunflower-seed products. After sunflower-oil prices collapsed during the 1975 growing season, the mill lost approximately $14 million and found itself obligated by contract to distribute a sum greater than its own net worth to 2500 farmers. The excursion into sunflower seeds as a source for oil nearly threw the mill into bankruptcy. Despite this setback, the oil mill survived because of strong membership support. Today, PCOM serves 85 member gins representing about 17,000 producers. Anyone driving past the PCOM complex sees mountains of cottonseed awaiting processing. The complex receives about 1200 *tons* of cottonseed per day. In 1987, it received 535,000 tons. That is cottonseed equaling eight of the *Queen Elizabeth 2*, the largest luxury liner in the world.

Throughout its history, the oil mill has been more than just a market for South Plains cottonseed. Particularly during the tenure of Roy Davis, management at the mill aggressively promoted the formation of cooperative gins and was a major stimulus to the strength of the cooperative movement on the South Plains. If farmers raised 25 percent of the cost for a new gin, the oil mill matched the funds, and the Houston Bank for Cooperatives matched the combined total to get the gin started. The oil mill's role in the cooperative movement had two important ramifications. The first was the idea for Farmers Cooperative Compress (FCC) at Lubbock to store South Plains cotton.

Farmers Cooperative Compress. Dan Martin explains the role of the compress. "Back in [the early 1950s], compress really meant 'compress.' The gins would only bring the bales of cotton to what they called a 'gin flat' density size [12 pounds per cubic foot] and then bring them to the compress to be stored and shipped when they were sold. But some of the cotton, that going overseas for example, had to be compressed down to a smaller size [32 pounds per cubic foot] to take up less space so the cost of shipping it would be lower. Back then, they called them compresses because many of the bales that were hauled in by the gins had to be compressed to a smaller size before shipment.

"In recent years," Martin adds, "they have been putting U.D. [universal density] compresses in the cooperative gins, and the gin itself is bringing the bale down to the right size." Universal density, now the standard size for cotton bales, is 28 pounds per cubic foot. "So," Martin continues, "80 percent or more of what goes to Farmers Cooperative Compress now is ready for shipment. Now compresses are almost strictly warehouses. When buyers buy and sell this cotton, they turn in a shipping order to the compress and it's [the compress's responsibility to get the cotton loaded on a rail car or truck and sent to its destination."

The idea for Farmers Cooperative Compress originated during board meetings at the oil mill. In 1948, one year after the boom that put Lubbock on the cotton map, FCC began operation. Prior to the establishment of FCC, facilities around Lubbock to compress and store farmers' cotton were highly inadequate and some gins had to haul their cotton long distances to obtain these services. Many growers unable to get warehouse receipts sold their cotton for $10 to $20 a bale below the loan price. These and other conditions led the farmers to set up their own compress. FCC pioneered innovations such as weighing and sampling the cotton on an automatic system at the compress and giving fast, accurate service in returning the warehouse receipt to the gins. By 1965, the compress had 84 warehouses and storage capacity for 600,000 bales. Today, it can store over one million bales.

The second major ramification of the oil mill was that information developed for the board of directors at the mill led to the organization in 1953 of the Plains Cotton Cooperative Association (PCCA), which ultimately became the marketing arm for cooperative producers in Texas and Oklahoma.

Plains Cotton Cooperative Association. PCCA quickly became a powerhouse in cooperative agriculture. In 1963, a year when 18 cotton firms closed their doors in Lubbock, PCCA opened compresses in Sweetwater, Texas, and Altus, Oklahoma. Membership totalled nearly 12,000 patrons. PCCA marketed 80 percent of the 1.2 million bales of cotton ginned through 148 farmer-owned facilities. More than a million bales of cotton were assembled at four cooperative compresses, the two owned by PCCA at Sweetwater and Altus, and the non-PCCA Farmers Cooperative Compress in Lubbock and Plainview Cooperative Compress at Plainview, Texas. Through his cooperative, the producer controlled the ginning, transporting, storing, compressing, and marketing of his cotton.

Roy Davis didn't create PCCA, but he was the "midwife" who delivered the organization into the world. Dan Martin recalls, "Five years after Farmers Cooperative Compress was organized through Roy Davis's help and [that of] the members of the oil mill, they organized PCCA.

"Farmers in the early 1950s were trying to solve a problem they had always had," Martin explains, "and that was fluctuation in the marketplace. There was a favorite saying, that 'When cotton markets are good, buyers are running all over you; when cotton markets are bad, you can't find a buyer with a search warrant.'"

Plains farmers knew they had to find a better way to market their cotton. A U.S. Department of Agriculture study showed higher marketing costs at Lubbock than at 11 other locations studied. Because the marketing effort in the Lubbock area was fragmented, with each gin and producer essentially doing their own, Lubbock's costs were $14.20 per bale as compared to $12.40 for Dallas, $10.40 for Memphis, and $6.25 for Augusta, Georgia. If that wasn't bad enough, another study showed Lubbock's costs had risen much more rapidly than those in other areas over the previous eight years.

Farmers also needed a cash market every day, not just when the market was strong. And they wanted their cotton marketed more aggressively. "The farmers," Martin recalls, "had been saying, 'Why can't we organize a marketing association to market or find a market for our members' cotton every day instead of spasmodically, as markets go up and down?'" A committee of farmers spent several weeks studying the situation. Much talk and planning went on in the oil mill's board meetings. As was so often the case, Roy Davis found a way to get the job done.

"Mr. Roy," says Martin, "had about $12,000 in some checks that had been written to members for retirement of oil mill stock or dividends and they couldn't locate the members. The checks had never been cashed. They were just lying over [at the mill], which meant the money was available at the bank. Mr. Roy stepped forward. He said 'Plains Coop Oil Mill has $12,000 worth of checks. I will let you have this money with the provision that, if anyone ever comes back and wants to cash their check, you have to do it for them.' That was the seed money that started PCCA."

Vision seemed a Davis family trait. That PCCA was founded at all is largely due to Roy Davis. That it became a progressive, farsighted organization is largely due to his son, Dan Davis, its former manager.

Dan Davis – Visionary Son

The younger Davis is himself a pioneer in the field of agriculture. As PCCA manager, he fomented the electronics revolution in the cotton-textile industry, which resulted in instrument classing of cotton and in development of TELCOT, a communications system for bringing the cotton market to farmers at their own gins.

Born in 1930, Dan Davis spent his early years in Plainview, Texas. The family lived in Houston when young Dan was between the ages of six and 13 and Roy Davis worked at the Bank for Cooperatives. When the elder Davis took charge of the oil mill in 1943, the family went to Lubbock. For Dan, it was a new home; for his parents, a return to a former home. Dan attended Texas A&M University, where he earned a bachelor's degree in business administration. He completed the Air Force ROTC program at A&M, and served on active duty as an Air Force officer from 1952 to 1954. After a few years in public relations work, including production of a weekly television program in Lubbock, he applied for the manager's position at PCCA. In June, 1956, he was hired.

Just a decade later, a magazine article described the effect of PCCA under Dan Davis on cooperative cotton farming in West Texas. It said, "Dan Davis and others on the [PCCA] management team have not been content to simply follow along after other cooperative leaders. They have ventured boldly into new ways, but only after sound analysis of their chances for success.

"The results have been good. In addition to the cash returns . . . farmers receive other benefits — having PCCA as their spokesman in industry and congressional hearings and with government agencies, as one example."

One PCCA member and cooperative leader estimated that the cooperative complex around Lubbock improved the income of cotton farmers on the Plains by an estimated $76 million from 1957 to 1963. Even independent farmers in the area were thought to have benefited through better prices the cooperative achieved. Indeed, if you want to know "Why cooperatives?" ask Lubbock.

SCIENCE IN THE COTTON PATCH

High Volume Instrumentation - Controlling Quality

When Dan Davis started work at PCCA, instruments for evaluating cotton were not new. Though primitive by today's standards, they were in development in laboratories at Texas Tech and other places and in use in some textile mills around the country. It was at PCCA, however, that accurate instrument testing and classification of high volumes of cotton moved from research to application.

Jim Parker recalls, "Dan was a very important figure at that time. He was instrumental in doing all this. His father was pushing him," he adds, "but Mr. Roy didn't know anything about the technology. Dan knew more about the value of the cotton and had a good feel for the use of computers and instrumentation."

These developments were essential for the survival of the West Texas cotton industry. Dan Davis explains why this was true. "Cotton from our area had a horrible reputation in the textile industry and, unfortunately, most of it was accurately perceived. We had absolutely horrible cotton. It was short and the fiber was weak. It was lousy cotton, and it was very difficult to get a handle on how to improve it." That fact did not discourage Davis and the staff at PCCA who were determined to find approaches to overcome the problem. "One," Davis says, "was to develop instrument evaluation so that we could accurately tell the strength of the fiber, its fineness, color, trash content, and so forth. The classing system and marketing system was [up to that point] so poor that, if a producer raised better quality than his neighbors, then he didn't get any more money for it. The classification system [most of which was still done by hand] was not sensitive enough to detect the fact that he produced fiber that was deserving of a better price."

Dan Davis is short and stocky. Under snow-white hair is a round, wide, open face that holds a pair of intense blue eyes that burn cold with determination. He leans forward in his chair and describes the situation he and PCCA faced. "You can imagine your frustration if you are responsible for selling the cotton production for 10,000 farmers, and nobody would buy your stuff because it is absolutely lousy merchandise. We were sitting there selling cotton for cents a pound less than cotton farmers in the mid-south or far west."

Emerson Tucker, the chief engineer of PCCA, well remembers those days. Tucker is a Connecticut yankee who came to West Texas for an education at Texas Tech and returned for a career, working closely with Dan Davis to help translate ideas into technical achievements. "A lot of people tested samples," Tucker says, "but the goal of PCCA was to develop techniques for testing all the bales in a shipment of cotton instead of a select few," with what now is called high-volume instrumentation or HVI. "Some people might test for strength only 10 percent of the bales in a shipment, but we were looking at 100 percent testing."

Another member of the team was Joel Hembree, who came to PCCA in August, 1964. Hembree had graduated from Texas A&M University in 1926, earned a master's degree in agricultural administration, and enjoyed a distinguished career in agriculture even before joining PCCA. He had worked with the Cotton Division and Commodity Exchange Authority of the USDA and was professor of marketing and statistics at the University of Arkansas Research Bureau. And for about 12 years he had been director of cotton economics research at the University of Texas. Hembree's work at PCCA was in crop statistics and laboratory and instrument development. Working with the young Davis and Tucker, he assembled data on the varieties of cotton planted, the types of cotton grown, and the size of the crop. He then would estimate the quality of the various combinations of grade, staple, and micronaire of cotton expected to be grown, and the amount to be received by PCCA. He also worked with the laboratory and mills when they had problems involving PCCA cotton.

Davis also had formed a friendship with a man in Dallas named Glenn Witts. Witts was founder and head of the firm called

Motion Control. To Tucker, Witts was "one of the 10 smartest people I ever knew." To Davis, he was "one of the few true geniuses I ever met. He could do anything with dynamics or hydraulics or electronics, and he signed on to develop the equipment. We supplied the money and the guidance as to what was needed."

The arrangement between PCCA and Motion Control was fortuitous for both. PCCA needed Glenn Witts's genius, and Glen Witts needed PCCA's willingness to put his equipment to use. In Jim Parker's memory, Glenn Witts was less in need of the money he received from PCCA than he was of the professional gratification of having his equipment put to work. "Glenn was a fiddler, a tinkerer," Parker recalls. "He loved to get the soldering iron and a whole bunch of resistors and capacitors and solder them all together and make something. He wasn't a very good manager, though, and couldn't have cared less about [public relations] or sales." Witts had been rebuffed repeatedly by prospective buyers. But he worked closely with PCCA for several years to develop high-speed testing lines for determining fiber length, strength, micronaire, trash content, length uniformity, and color for each bale. All these are characteristics affecting the quality of cotton in the bale and the cooperative's ability to sell it for a good price. Together, Motion Control and PCCA confirmed that solid state electronics could reduce the time required for instrument tests. PCCA pushed for an instrument line that would reduce test time from five minutes to five or 10 seconds. It would be used on a high-volume basis to provide instantaneous results on each bale of cotton during the cottonmarketing season. The purpose was to develop a technical profile on each bale, which then would be used to market the product to mills.

Two of the principal tests are color measurements with the colorimeter and micronaire readings by an air-flow instrument. The colorimeter enables the cotton classer to make an objective check on color, one of the important factors which, together with "leaf" (the amount of trash) and "preparation" (ginning), determine how a sample will grade. An air-flow instrument is used to determine fiber fineness or maturity, which cannot be measured accurately by a human classer. It subjects a cotton sample to standard air pressure, and the amount of air flow is indicated on the machine as a micronaire, or "mic," reading. A fine-fibered, immature cotton gives a low reading, whereas a coarse and mature sample results in a higher measurement. The innovations at

PCCA had a ripple effect on cotton marketing. Mills traditionally used a micronaire test that measured fiber fineness to determine spinning characteristics and dye acceptability for cotton. Having initiated its own micronaire testing, PCCA could fill orders from its stocks when mills specified what "mic" cotton they needed.

Emerson Tucker recalls that as early as 1960, when he joined PCCA, the association had a micronaire testing line, a colorimeter, a leafer, and a condition checker. "We put a value on all the bales and I think we tested 437,000 bales in this format," he says. "We had come up with a relatively good instrument line. Not perfect but pretty good."

Producers in West Texas had been penalized on a regional basis because mills assumed all cotton originating on the Plains was low micronaire, fine fiber. Now PCCA was using instruments to identify the superior bales and overcome some of the prejudice. Within a few years, independent ginners on the South Plains requested micronaire testing, and in 1964 the USDA classing office at Lubbock, the largest in the world, began offering micronaire values in its classing procedures.

By 1967, three of the first five instrument lines developed were in operation in Lubbock, one at the Textile Research Center and two at PCCA. The TRC got its first unit in a roundabout way. An independent cotton marketer had purchased one of the first HVI units produced by Motion Control and planned to start doing the same type of testing and analysis PCCA had begun. He was killed in an accident shortly after that, and his wife donated his unit to the TRC. The TRC later would pick up units that were discarded by textile manufacturing corporations.

A 1967 article in the PCCA newsletter discussed averages in cotton characteristics for the Lubbock area, an analysis in detail that would not have been possible just a few years earlier. It noted, "A sharp upswing in quality of cotton from the 1967 crop, as the High Plains, long characterized — and criticized — as a land of cotton, much of poor quality, took great strides in quality improvement." The results of the tests showed Lubbock-area cotton moving toward acceptable standards in the industry. This was a result of better seed, better production practices, and instrument testing that scientifically proved it wasn't all quite as bad as it had been portrayed. Later tests would show it was, in fact, far superior to other strains of cotton in certain applications. "USDA cotton quality reports," said the newsletter, "show average staple for the '67 crop was 31.3 thirty-seconds of an inch for the Lubbock area. . . . Other quality factors were evident in the cotton produced in PCCA operating areas. Lubbock cotton averaged a micronaire of 3.2 – not as good as farmers had hoped, but the cool season contributed to the low average readings. Desirable 'mike' [*sic*] range . . . is 3.5 to 4.9.

"Fiber strength, now a sought after quality factor, is preferred at 80,000 pounds [per square inch, Pressley] and above. Average fiber strength for Lubbock [was] 79,000 pounds." (The per-square-inch, Pressley measurement indicates the breaking strength of one square inch of cotton fibers and has since been replaced by a "grams per tex" measurement.)

By the 1980 season, the first USDA classing office relying on instrumentation opened in Lamesa after years of testing in Lubbock. Instrument testing was a valuable breakthrough for West Texas cotton. Instruments delivered readings based on the properties of cotton fiber and not on the basis of where the cotton originated. Furthermore, instruments identified properties human sight and touch could not determine, and in so doing brought the textile industry into the twentieth century. Cotton producers had not been able to supply as much technical data on the spinning qualities of their product as the synthetics industry had been doing with man-made fibers, but the instruments brought them much closer than ever before.

Instrument classing was not appreciated in every corner of the industry. To buyers, it was an athema. As Jim Parker describes it, "Cotton buyers saw these instruments as a means of just wiping them out. If you're going to measure the cotton on instruments, why do you need a cotton buyer?" The ability of grower-members of cooperatives to test and classify every bale of cotton, then sell it directly to textile mills by meeting their specifications, would literally cut the "middle man" out of the picture.

Parker describes an incident that took place in the early 1970s. "Burlington Industries in Greenville, South Carolina had bought two of these instruments from Motion Control to test and evaluate all the cotton they bought. They had an internal fight over these things," he says. "I mean it was really a fight. These [buyers] were old, old-fashioned. They were doing the same thing they had been doing for 100 years. It so happened that at Burlington a cotton buyer was an officer of the company and rather powerful within the company." The man was a long-time friend of Parker. Parker, characteristically frank, describes him as "old and stubborn.

"I wasn't at the meeting where they had the fight, but I heard later he told [the company's] board that if they went that route, they were going to shut the company down because the instruments didn't measure the fiber properly, and there was no way to really take the results and apply them to the different yarns and fabrics. He said the mill was going to collapse," Parker relates, "and he wasn't going to be responsible for it. He said, 'You people are on your own. If you go bankrupt it's your fault.' It scared them to death. That's what I heard. They backed off. They gave us one [instrument] line, and they stored the other one somewhere.

"About two or three years later, my friend retired and one of the manufacturing vice-presidents was visiting us at TRC. We just happened to be running and testing cotton on his old instrument line that day. He stood there and watched and said, 'You know, we made a terrible mistake. This is the only way to go. Why did we do what we did? It was awful.'" Parker adds, "They got back into the business right quick. I think they bought new lines and equipment. Burlington [tests cotton] now, and every major company now either has their own HVI equipment, or they get somebody to test it for them."

As late as 1980, W. R. (Bill) Moore, then president of the Texas Cotton Association, railed at instrument classing. At the annual meeting of his organization, he thumped the USDA instrument classing program and Jesse Moore (no relation), the man in charge of it. "USDA wants a machine to do it all," he said. "If Mr. [Jesse] Moore finds administering machines easier than administering people, then he should request a job transfer. The job Mr. Jesse Moore occupies was formerly graced by men who could accept the challenge of grading fairly the entire American crop and on the balance do a good job. The job is now graced with a new broom who wants to sweep away not the dirt but a world system of cotton classification that today is over sixty million bales annually. Talk about Big Brother. We have got to fight instrument classing, gentlemen, at the grassroots level," he implored. "The main losers will be the producers. Go home and tell your producer friends to let everyone they know, with any clout, to stop this man and others in the USDA from this move against Texas cotton farmers. I think, if this program flies, it will go down in Texas cotton history as a change that cost our producers millions and millions of dollars." If Bill and Jesse Moore had been related, they would be distant relatives.

Jim Parker admires Jesse Moore. He says, "I wouldn't want his job. It's a horrible position." And he credits him with working hard to bring the USDA system to the level of accuracy it now enjoys. Today, the howls against instrument classing are only faint echoes from the past. Parker sees no retreat from the march of technology and the benefits it has brought to the cotton and textile industries.

TELCOT – Opening Markets

During PCCA's early years, cotton producers had no organized system to market their crops. The only way to determine the value of the crop was the price offered that day by a local buyer. Little information was available for producers to anticipate the value of their cotton a few days later, nor were they aware of prices being offered for cotton "a few miles down the road." This situation was compounded by access to only a few buyers. Competitive bidding for cotton was still a thing of the future, and there were no guarantees producers would receive payment once their cotton was sold.

Similarly, buyers had to rely on a repetitive, labor-intensive system to acquire cotton. Often, the system was flawed and subject to error. Buyers had no way of knowing prices being offered by their competitors and, in some cases, may have offered prices in excess of market value. Access to cotton was limited and expensive to locate. To overcome this, buyers employed country agents, or "road runners," who traveled to local gins to review available cotton and negotiate the sale. The process often was preceded by quality recaps prepared by local gins and relayed to buyers via telephone. The system required numerous middlemen and made cotton marketing an expensive proposition.

The system was slow and time-consuming for local gins as well. Manual recapping of cotton qualities required intensive labor. This process involved a gin clerk who used an adding machine to figure the recap. Information was relayed back and forth between the producer and prospective buyer through the gin office until a sale was consummated or the offer rejected. Often the producer and buyer could not agree on a price, and a few days or hours later the process was repeated. The element of human error was an ever-present possibility, but the system was a "necessary evil" because no alternative existed.

By the late 1960s and early 1970s, the concept of a centralized clearing house to acquire PCCA members' cotton swiftly and efficiently, then re-offer it simultaneously to numerous merchants for competitive bidding, was being born. As with HVI, Dan Davis was its champion. His goal was to use the latest in computer technology to create an electronic marketing system for PCCA cotton. TELCOT became a commercial reality in 1975, when remote terminals were installed in 15 cotton buyers' offices in Lubbock, Dallas, and Memphis. The concept of TELCOT is similar to that of the New York Stock Exchange. The producer is provided a number of options to sell his cotton, access to buyers, and the opportunity to receive the most competitive price available at the time he wants to sell his cotton. More importantly, TELCOT guarantees each producer will receive payment for his cotton. Through TELCOT, buyers are able to shop for the specific kinds of cotton they need, in the amounts they need, at the best price available at the time. The cooperative gin consummates the sale, through the network, on behalf of the producer-member. The sales instructions then are sent on to the compress so it can locate and ship the bales.

Today, TELCOT terminals are in over 160 PCCA-member gins in Texas and Oklahoma, as well as in the offices of more than 40 major buyers of U.S. cotton. The producer simply stops at his or her gin office to get information or conduct transactions. Sales and purchase information is maintained and updated in seconds. In addition to keeping subscribers fully aware of marketing data and developments in cotton around the world, the system also provides information on other selected crops and general agricultural news from national and local sources. It also serves as an electronic bulletin board for communication within the cooperative and between its members. PCCA calls the system the farmer's "Window of the Marketplace," and with good reason.

BLUEPRINT FOR TEXTILES

In June, 1971, Lockwood Greene Engineering, a large national firm specializing in design of textile mills, was authorized by the Texas Industrial Commission to conduct a study on the potential for textile production in the state. The firm, which presented its extensive findings nearly a year later, was asked to answer the question: "Do conditions in Texas appear favorable for the textile industry to the degree that further study and investigation by the Texas Industrial Commission, other State agencies, educational institutions, or private investors, are warranted?"

In its path to the answer, the study first observed that the national textile industry was highly concentrated at the time in four states:

North Carolina, South Carolina, Georgia, and Alabama. It cited U.S. Department of Commerce data showing 91 percent of the spinning spindles in the country in those four states, where they produced more than two-thirds of all cotton fabric woven in the U.S. Texas did not fare well by comparison. "The balance of the industry," the report found, "is somewhat scattered through New England, the Middle Atlantic States, a minor portion in the Midwest, and somewhat isolated segments in western states such as Texas and California."

But the study also found that southeastern dominance was fast eroding. The textile industry in the region was being cannibalized by other, more technologically advanced industries. "In the decades of the 50's and 60's," the report said, "these four states have been among the most active in the nation in spurring industrial development. These efforts have been extremely successful, particularly in North Carolina and South Carolina." This "general industrialization" was creating major problems for the textile industry there, just as it had done in New England before the industry migrated to the Southeast. "The general labor pool formerly monopolized by the textile industry now must supply the entire range of growing manufacturing employment in these states. Wages are rising sharply. Skilled employees and semiskilled employees . . . have left the textile industry for higher wages in other types of manufacturing.

"In the recent past, significant problems of absenteeism, lower productivity and turnover have appeared in many textile mills. The resulting increased costs cannot entirely be passed along to the purchasers of the textile industry's products because foreign competition is more aggressive than ever before.

"In short," the report said, "the present U.S. textile industry, largely concentrated in the Southeast, must find additional ways to reduce its manufacturing and operating costs if it is to continue to exist." The textile industry in the Southeast was becoming not only noncompetitive but threatened. The study then focused on Texas. "The history of the textile industry in Texas is uneven. Some of the mills in Texas are extremely small; some are large. There have been a number of significant failures of past textile ventures. Some of the present mills are extremely profitable."

U.S. government data showed that less than one percent of U.S. spindles were located in Texas and that they wove just 1.5 percent of the fabric produced in the entire nation. Furthermore, although the state produced 3,198,000 bales of cotton in the 1969–70 season, only 5.47 percent of it was sold to mills in Texas; the rest was exported from the state, . . . despite the fact that Texas was considered a major apparel-manufacturing state.

When the data on cotton production and apparel production were put together, they demonstrated the most significant finding of the report and the supreme irony of the Texas agri-industrial climate. Cotton was grown in Texas. Apparel was made in Texas. But the middle part of the process, turning cotton into textiles, was not done in Texas. Cotton was shipped out of state and later shipped back in the form of fabric. Texans were buying back their cotton and paying handsomely for the privilege. Roy Davis and others were painfully aware this was happening. They had plans to remedy the problem but needed support. A solid work of analysis by a prestigious firm, the Lockwood Greene study would provide that support. It would help demonstrate to others what the Davises and other visionaries knew all along to be true.

Lockwood Greene studied the entire state in analyzing regional potential for textile production. They considered labor and land requirements, utilities, transportation needs, waste disposal, water supply, taxes, insurance, construction, and other costs. They settled on three regions where the potential seemed greatest. One was Central Texas, from around Austin to San Antonio. Another was the Rio Grande Valley. And the third was the High Plains and Rolling Plains of West Texas. The firm answered the question posed by the Texas Industrial Commission, and the answer was affirmative: ". . . the potential for development of the textile industry in Texas appears high and . . . further efforts towards the realization of this potential are warranted." Its report was a blueprint for Texas to move further into the textile field. And one of the areas to which it pointed was West Texas, Roy Davis's own backyard.

Roy Davis was inconspicuously named as one of the contacts made by the Texas Industrial Commission for the Lockwood Greene study. There is a strong suspicion on the part of many today that Roy Davis was more than just a "contact" in the study. He may have been the one person, more than any other, who was responsible for having the state government of Texas commission the study. It was just after the Lockwood Greene study was completed that Roy Davis came to Texas Tech University to work side by side with Jim Parker. Dr. John R. Bradford, dean of the College of Engineering and director of the Textile Research Center, made the announcement. He said Davis would be "making recommendations concerning the needs of the textile industry and of Texas producers of natural fibers relative to the research efforts of the center." Bradford said the appointment resulted from a recommendation made by the Textile Research Center's Advisory Committee, headed by farmer-rancher Don Anderson of Crosbyton, Texas.

Anderson spoke of the stature the Davis appointment would bring to the research center. "Roy Davis has given a lifetime of interest and effort to fiber producers, and the committee feels that the knowledge gained in that lifetime is an invaluable asset which Davis is bringing to his new office within the Textile Research Center."

In talking about the elder Davis's influence, Dan Martin recalls, "his dream just kept growing. First, it was that the cooperatives needed to be in the seed business, then they needed to be in the warehouse and compress business, then they needed to be in the marketing business. And then they needed to be in the textile business. More than once," Martin adds, "I heard him say, 'We need to bring the textile mill to the cotton patch.'"

Not everyone agrees that Roy Davis's dream was to get "cooperatives . . . in the textile business." Bill Blackledge, an attorney who was PCCA's corporate secretary, maintains "Roy Davis thought that a textile mill ought to be built [on the Plains, but] . . . he didn't think that the cooperative ought to do it." Blackledge acknowledges that he isn't "speaking from first hand information," but says, "I heard some comments made between Dan and his dad that indicated to me that his dad was totally against the cooperative getting into the denim manufacturing business."

According to Blackledge, Roy Davis "wasn't violently opposed to it . . . he just felt like it was getting out into dangerous territory for us to think in terms of the monstrous investment that's involved in a textile mill being owned by thousands of farmers. How it could be done was just a little beyond his imagination. "Roy's idea," he says, "of running a coop was to have one man on the board of directors from every gin. So he had an 86-man board running the oil mill. We had a 160-man board running [PCCA]. It was a totally unworkable management style for a coop, because how do 160 directors make a decision? It's like running a Greek city-state. Everybody gets a vote." Blackledge feels that Roy Davis could not have reconciled the construction and operation of a textile mill with his management style.

"No," he adds, "Roy thought there ought to be a textile venture in Texas . . . and he was instrumental in getting this Rural Development Act passed through Congress. He was a good politician. He had lots of friends in Washington. But actually, it was Dan Davis's idea [for the cooperative] to build that denim mill." In Blackledge's opinion, Dan Davis never has gotten the credit he deserves for that idea, nor is he ever likely to.

In fact, Dan Davis's and PCCA's flirtation with fabric manufacturing far predated the Lockwood Greene study. "Earlier," Davis recalls, "PCCA had tried to get into the textile business in Asia. This was during the mid-sixties, when Orville Freeman was [U.S.] secretary of agriculture and Dorothy Jacobson was his assistant secretary for foreign affairs. . . . Dorothy had asked me to see what might be done to activate Section 407 of Public Law 480, which provided for private trade entities." The legislation, administered by the Agency for International Development, an arm of the U.S. Department of State, encouraged private American corporations to become involved in economic development efforts in the Third World.

Davis continues, "At that time, we could see tremendous textile expansion that was taking place in Asia and we were anxious to have joint ventures in which U.S. cotton producers would have a 50 percent equity position as these plants were being constructed at a rapid rate in many of the countries in the Far East. We had set up a company in Hong Kong called Amerasia as a holding company, and then we had a company in Thailand which was Thaiamerican Textiles, which was 50 percent owned by the Thai partner and 50 percent by PCCA. We got a $75,000 grant from the Agency for International Development to do a feasibility study over there.

"The mill would have been a $16 million plant. We got land just beyond the airport in Bangkok, and they had the thing cleared, letters of credit, and selected the machinery, but the cooperative didn't have much money." To finance the plant, the U.S. Government would lend PCCA cotton instead of money, 20,000 bales from the government's stores for which farmers already had been paid. "We could sell the cotton," says Davis, "take the proceeds, build the mill, and then pay for the cotton over a 20-year period. After a lot of time and money spent over there to get it going, the Democrats lost the [1968 presidential] election and one of the first things the Republican administration did was to torpedo our textile venture. And so the Far Eastern textile thing never was completed," Davis says, adding a postscript that would be familiar to American business in the 1970s and 1980s. "The Japanese took our half interest in the mill and built a plant which paid for itself in 18 months."

When asked about his assertion that the incoming Republican administration killed the textile venture, he doesn't shy away. He claims they targeted this particular project, and they did it under pressure from the domestic textile industry. "The textile interests were against it," he explains, "even though we were not going to produce textiles for importation into the United States."

Dan Davis's political thinking may be partisan. His family political tradition was unwaveringly Democratic. Roy Davis told Jim Parker that he always, in *every* election, voted for the candidate of the Democratic Party. But this foreign venture in the mid-1960s shows that Dan Davis was no parochial thinker when it came to marketing cotton. He repeatedly proved far ahead of his time. Whether his claim about political interference in the Amerasia venture is true or not, PCCA would do better at working the political system in the future.

Experiment at Crosbyton

In the early 1970s, the attention and efforts of Roy Davis, Dan Davis, and the PCCA staff were focused on three major elements of the West Texas cotton culture. The first was HVI, the use of instrumentation and other technical advances to improve the quality of the crop and its attractiveness to potential buyers. The second was servicing existing markets and building new ones, domestically and abroad. This included the use of TELCOT, the new technology through which cotton could be marketed. The third and most important was improving the overall ability of the cotton growers to control the marketing and profitability of their crops. That is where American Cotton Growers enters the picture.

American Cotton Growers (ACG), whose story this is, existed as an idea for many years, its roles and objectives seen only in the minds of a few. It did not suddenly emerge as a distinct entity, rather as a slowly evolving concept designed to fit both the unwritten "master plan" of the visionaries and the exigencies of the time. As biblical fathers begat generations of sons, West Texas cooperatives begat other cooperatives, all in the same family but each with its own features and contribution to the whole. ACG's evolution was complicated, and a brief overview may be helpful.

PCCA, under the leadership of Dan Davis, developed the concept of the "super-gin" and encouraged three ginning communities — Crosbyton, Wake, and McAdoo — to build one. The super-gin, completed in 1973, was located at Crosbyton, and the cooperative association that built and operated it was called American Cotton Growers. ACG also ran its own cotton marketing pool. When PCCA wanted to build a denim mill and could not, a decision was made that ACG would build it instead. ACG expanded to include 27 ginning communities. The original organization of three communities that built the super-gin became the Crosbyton Gin Division (CGD) of American Cotton Growers.

A MATTER OF SURVIVAL

Dan Davis recalls the purpose of ACG. "[It] was really organized to try to correct some of the problems we'd encountered in other cooperatives here in the State of Texas. We created a new cooperative to try to build from the ground up a structure that would have a substantial impact on the net income of its producer members."

As a practical matter, ACG was organized to build a centralized ginning facility, a so-called super-gin. According to Dan Martin, Dan Davis steadfastly maintained that it made no sense to have "a bunch of little, 5000-bale gins scattered around. You needed a gin big enough to gin 20,000 bales, [with volume enough] to extend your ginning season from a three-month period to a six-month period."

L. C. Unfred adds, "It takes about 7000 or 8000 bales just to break even at a gin location. If you can gin 14,000 or 20,000 bales, then you can make some money and return a nice dividend of maybe $10 or $20 a bale to [the producers]." The idea made good economic sense; ginning, like many other processes, is one that improves and costs less with increased volume.

A large investment was required to build a modern gin with a universal density press. By the early 1970s, gins were buying and installing their own U.D. presses, performing the compressing function before the cotton went to the compress. In the words of Bill Blackledge, "It cost something like a half-million dollars to get a universal density press at the gin. Well, you didn't have enough volume in each one of those little gins out there to justify that kind of capital investment. It meant organizing an . . . American Cotton Growers which would . . . take the whole job under its wing in different divisions and . . . could afford to spend the money out at the gin in order to make the system more efficient."

Dan Davis also says of ACG that, "One of the initial thrusts of it was to set up a marketing pool, with the membership of the pool to be concentrated in certain communities."

According to Martin, Davis "integrated the super-gin concept with a pool marketing concept. He hit [the farmers] with two ideas instead of one." The "pool" concept is basic to cooperatives. Farming is fraught with dangers over which man has little or no control. The most severe of these are natural calamities and uncertainties of the marketplace. Either can cripple entire segments of the industry. A drought, flood, freeze, or high winds can destroy all the crops in a geographic area. A plunge in demand for a specific commodity can bankrupt producers of that commodity around the world. A few bad crops in succession can wipe out a family farm. There is little farmers can do individually about the risks. But there is a great deal they can do together to make their own opportunities and try to assure survival and achieve prosperity in the marketplace. Farmers who are member-owners of cooperatives are able to share risks, opportunities, and resources.

The cooperative spirit that came to exist on the American frontier was born partly of goodwill, but more of necessity. People cooperated, or they might not survive. This meant sharing time, effort, and resources for the common good. Dan Martin describes the philosophy. "As long as we have to share the risks together, let's share the means of survival." That is the essence of the cooperative spirit. But the farmer also is the archetypical "rampant individualist." Because they take so many risks in their work and survive by their own labors, farmers tend to be very independent. They simply do not like being told what they can or cannot do. This dichotomy of attitudes would be seen repeatedly throughout the history of American Cotton Growers.

Martin explains why it made sense for farmers to enter a marketing pool. "When they don't belong to a pool, farmers sell their cotton during a limited time of the year, the period just during and after ginning. Marketing through a pool, however, they are selling throughout the year on all markets. . . . The pool will never hit the high of the market, but neither will it hit the low. Everybody will get a good, high average, as opposed to a guy who makes all his own marketing decisions. If he is smart enough to figure out when the high of the market is and sell his cotton then, it's great. But most of them, will wait too long and can't catch the falling market. So one year they might sell on the high, the next year on the low."

Martin illustrates. "In 1973 . . .the first year of operation of American Cotton Growers at Crosbyton, before the denim mill was built, their cotton average 'bring' was about 63 cents a pound. Well, in '73, the market was way up. We had farmers who sold their cotton for 78 cents a pound. But we had a bunch of them who chased it all the way down to 38 cents. For the total economy of that community, it was a lot better for all of them to get about 63 cents a pound than for some to get a bunch and others to hit bottom.''

Dallas Brewer recalls being caught in the reverse of the situation Dan Martin describes, just as the cotton market began to go up. ''We contracted too early [in 1973] and cotton just kept going up. I had mine contracted for 34 cents a pound. And in the fall, guys who came into the gin who didn't contract were selling their cotton for 74 cents and 75 cents.''

If pooling makes so much sense, why would farmers ever balk at the notion? Dan Davis had 17 years of experience with the PCCA pool before ACG was founded. He attributes early pool problems of the cooperative to a combination of farmers who weren't committed to the concept and competitors who wanted to see it die. ''We'd had some real difficulties in making the pool concept work here in Texas. Producers were not as pool-minded as is the case in California and perhaps some other parts of the United States. And so at that time [late 1950s], we had just a few members in many different communities. Typically, the members of the pool were a minority, and it became very difficult to sustain their morale. In a lot of instances, the local board of directors and local management of cooperative gins were anti-pool. Certainly all of the merchandising competitors of the cooperative were anti-pool and did not want to see the thing get started. Much of the opposition centered around the country buyers, and we found that there was no very good yardstick with which to measure the success or lack of success of the pool in contrast to prices farmers received [from the country buyers] outside the pool.

''At first [in the 1950s and early 1960s], members were inclined to sell what would sell to our competitors, and then put the cotton that was not immediately marketable into the pool. It meant the deck was kind of stacked from the beginning against the pool, since we had the cotton to work with that nobody else wanted. After trying an unrestricted pool for a period of time,'' Davis continues, ''we formed two pools — one for producers who would sign up all of their production and give us the total crop, and the other for that portion of the membership that just wanted to pool the cotton which no one else would bid on.'' This arrangement, in which one faction participated completely and another at its pleasure, clearly violated the cooperative spirit. But there was no clear way of telling what system, or combination of systems, was better for the farmer.

''The 100 percent pool was successful,'' adds Davis, ''but we found that at the conclusion of the marketing season, whatever the pool outturn [earnings], the local merchant could always have done better [for the farmer] than the cooperative, had the cotton been entrusted to him . . . at least so [the merchant] told it. Then the government came along with the new cotton program in 1961, I believe, in which substantially all of the crop was sold to the government. The government appointed sales agents to sell the cotton. That kind of put the pool out of business, and we did not try to resurrect it until more than 10 years later, when American Cotton Growers was organized."

There were reasons for the resurrection of the pool in the early 1970s. "Instead of buying the cotton from farmers," Dan Martin explains, "the federal government began lending farmers money which would be repaid by them out of proceeds from the sale of their cotton. This still is the case today."

Dallas Brewer explains how the loan works. "The U.S. government puts out a 'loan card' that says how much it will loan against a certain type of cotton. For example, a middling grade with a one-inch staple might bring 50 cents a pound. Then, through the year, cotton buyers will tell you, 'Well, I'll pay you 2000 points over the government loan.' Each 100 points represents one cent. So, they would be offering you 70 cents per pound . . . the 50 cents in the government loan and the 20 cents above the loan. The farmer has some protection. Even if cotton buyers aren't paying anything over the loan, [the farmers] are guaranteed that price for that particular grade and staple. If you couldn't sell that pound of cotton," Brewer says, "or you had to sell it for 40 cents a pound, then you would just go ahead and put it in the government loan and get the 50 cents. You would turn the cotton over to the government, and they would market it."

"In the early sixties, most of the cotton sat in warehouses for long periods of time," Martin adds, "and when it did trade, it only traded at five to 10 dollars per bale equity . . . that is, five to 10 dollars per bale over the government loan. Since PCCA was buying, selling, and shipping about a million bales a year, they only had to borrow five to 10 million dollars from the bank to pay the farmer." That constituted PCCA's market risk, the amount for which they were responsible if they couldn't sell the cotton.

"What really caused the situation to change was in 1973," says Martin, "when cotton markets ran up as high as $150 a bale above the loan. PCCA had to borrow in the $80 million range from the Bank for Cooperatives to finance our buying, selling, and shipping of cotton. At the time, we only had a net worth of about eight or 10 million dollars. The bank really didn't notice this, I guess, until they laid it all on their books. They realized what they had done — lending $80 million against $8 million — and said, 'We'll never loan you that kind of money again. You don't have enough net worth to borrow [what you need] to market all your members' cotton. You've got to look for another way to market.'''

Cooperatives are supposed to be risk-sharing arrangements, but PCCA had put itself in the position of becoming a risk-bearing organization. Martin recalls that Dan Davis and C. L. Boggs, a young accountant who served as Davis's assistant, came up with the idea of pool marketing to get at least part of the risk off the cooperative and onto the farmer ". . . since you never advance more money to the farmer out of the pool than you know for sure you're going to get back." But the farmer still does not bear as much risk as he would without the pool, where one year could be very good but the next ruinous. The pool was one way of reducing the risk to PCCA. The other was by marketing through TELCOT. Martin says, ". . . with TELCOT, we could take what money we could have borrowed at the bank, based on our net worth, and market that much of our members' cotton. Over TELCOT, we could sell to other buyers, who normally would be our competitors, and the [result] would be that we would merchandise all our members' cotton."

One of the men heavily involved in the creation of American Cotton Growers was Jimmy Nail. Nail, a big man from Breckenridge, Texas, "didn't know what a cotton gin was" in his youth. But he knew that he wanted to work for a cooperative from the time he was a junior at Texas A&M University. "It was," he says, "my burning desire." He got his chance. Despite a master's degree in agricultural economics, Nail didn't exactly start at the top. "I ended up working for Dan's uncle, Weldon Martin, at O'Donnell [Coop Gin] on the scales at $1.25 an hour, 84 hours a week." That was in 1959. From there he went to South Gin, a farmer stock-owned gin, as manager. He was there until 1969 and helped turn that gin into a cooperative "two or three years after I arrived."

Dan Davis hired him for the PCCA staff in 1969. Around 1971, Davis put him to work on the concept of a "large, central gin that would gin cotton to be sold in a pool." Nail, along with Emerson Tucker and Wayne McElroy, PCCA's shop engineer, went to California and studied how large growers handled their cotton there. "We made films . . . of how they would rick cotton [a rick is a sled-like contraption moved along the turnrow, onto which cotton is placed], pick it up with large loaders, load it into either trailers or semi-trucks, haul it to a gin, and gin it."

PCCA's goal was not only to gin pool cotton in great quantity, but to raise the quality of the output. "We had been carrying on some experiments at Wells [Texas] to prove that cotton could be blended," Nail says. One of Dan Davis's objectives was to blend the cotton production of a large number of farms. He felt the level of quality and volume of sales would be raised if it were possible to blend cottons of varying characteristics and come up with a uniformly superior product.

Bill Blackledge talks about the 1971 Wells Project. "They used to haul [cotton] loose in a trailer. This was a project to stack the cotton in modules like they do today on the turnrow. One of the problems [when we had the smaller gins] was that the farmer took the crop off the stalk faster than the gins could gin it. The gin couldn't keep up with the harvest. Cotton piled up on the gin yard in the farmers' trailers, which were tied up. The farmer couldn't get the cotton out of the field," Blackledge says, "and it would get damaged on the stalk if there was bad weather during harvest season. What we were trying to do was stack the cotton in modules like they do today . . . and haul them away on the module truck." The module builder was important, according to Blackledge, because the old trailers "could only travel about 10 miles to a gin, so you had to have a lot of gins. One of the approaches we wanted to take was to expand the territory you could serve with a gin . . . to handle 35,000 bales a year. That would justify the capital investment it took to put a universal density press at the gin," but it could not be done using trailers.

Blackledge is an intelligent man, a deep thinker and stickler for detail, who seems as comfortable talking about the science of engineering as he does about the art of law. Emerson Tucker, a close friend, accuses him of being a "frustrated engineer"; he retorts by teasing the dialectical Tucker about being a "frustrated lawyer."

"We had a lot of meetings with gin communities, presenting these ideas," Jimmy Nail recalls. "Farmers at Crosbyton already were in the process of visiting with Wake about merging their two communities, so they were a ready-made audience. When it looked as though Wake and Crosbyton were going to go together, McAdoo wanted to go with them." The three ginning communities became the laboratory for testing the concepts of Roy Davis, Dan Davis, and other far-sighted cotton marketers and producers. As

Dan Martin puts it, "Dan Davis was always thinking about what is going to happen 10 years from now."

FINDING LEADERS

To make his ideas reality, Davis needed strong, effective leadership among the producers. "In Dan Davis's mind," says Dan Martin, "[there was] a very progressive bunch of farmers . . . in those three communities." The leader of the progressive group was T. W. Stockton. "It was T. W.," says Martin, "and he had some help," citing Compton (D. C.) Cornelius and Clyde Crausbay. "They had several strong leaders in those three communities, so Dan Davis went to their area to get the job done."

If American Cotton Growers had its origins in the vision of Roy and Dan Davis, it found much of its strength in T. W. Stockton. Bill Blackledge says, "T. W. Stockton at Crosbyton was one of the most effective men in putting together the growers' side of that first merger . . . we couldn't have done it without [him]. He was really instrumental in keeping everybody happy as far as the growers were concerned." Stockton had to be good to get the job done. As Blackledge says, "When you consolidate the gins, you take a gin out of a community. Then the banker gets unhappy, everyone gets unhappy, because you are taking a business out of that community."

If you were making a movie about the West Texas of 1973, or even 1873, and you told "central casting" to find someone to play the role of a community leader — weather-beaten and wise, firm and fair, stoic and strong — they would send you a T. W. Stockton. He does not talk much. When he does, he does so quietly and thoughtfully. He is widely respected in a land and livelihood where a man's honesty and dependability are the possessions by which he is most often judged. He is an epic figure in his place and time.

Stockton, later vice-chairman of the ACG Board, was born in 1924 on the very farm he operates today. In some ways, his story is typical of the "old lions" of the ACG Board. His father moved to Texas from Arkansas as a child, first to Abilene in 1906, then to Crosbyton in 1915. Stockton went to a local high school, and then, as was the case with so many of his generation, his life was interrupted by World War II. He was decorated for his role in combat. Returning home after the war, he attended college for a few years and, he says, "had an opportunity to go into farming."

That is what he did, starting with 160 acres and building it to about 3000 today.

TAKING ACTION

Stockton and the other local leaders could take the matter of consolidated ginning only so far. In order for the super-gin to be built at Crosbyton, the members of all three ginning associations would have to express their approval. On March 8, 1973, a special meeting was held to put the plans to a vote. One hundred and sixty common stockholders of Wake Cooperative Gin, Farmers Cooperative Gin at McAdoo, and Crosbyton Farmers Cooperative Gin convened to hear Dan Davis and his staff explain the advantages of modernizing the cooperative cotton marketing system. Davis had proposed the modernization be accomplished by organization of a new cooperative, which would be responsible for handling its members' cotton from the time it was harvested until it was delivered to the spinning mill. The new cooperative, to be called American Cotton Growers, would be formed by merger of the three associations into a new entity. The merger plan already had been considered and approved by the boards of directors of the three gin associations, and they had recommended their members adopt it.

Clyde Crausbay had "talked it up" in McAdoo, and the board there voted unanimously to join. "The timing was right," he recalls, "cotton was cheap. A lot of new rules and regulations made you want to do something."

Compton Cornelius of Wake, who brought in the support of his board, recalls the rules and regulations. The old gins not only were inefficient, they could not conform to the latest government standards. "It was the year that OSHA [the Occupational Safety and Health Administration of the U.S. Department of Labor] was getting rough." The old facilities could not meet the government's new health and safety rules. There were two gins in Crosbyton at the time, and Cornelius recalls they would have had to be rebuilt, or as he says it, "They were needing to get out of town . . . [the] gins were getting old and run down. We were having the same problem at Wake [Cooperative Gin]. We were having trouble with OSHA and getting rid of our burrs and trash, only we weren't in the town." He pauses and quietly adds, "It's just a dried-up community now."

The boards of directors of the gins had taken action. Now it was time for the members to speak. A resolution was introduced. It empowered the officers and directors of the proposed association, "to develop and participate in a reorganization plan for the modernization of the cooperative cotton system." The new cooperative, it said, ". . . shall be responsible for handling, transporting, ginning, storing, and selling its members' cotton; for the reduction of their off-farm costs; and for increasing their returns from cotton production." The resolution also proposed that a Gin Division Board be elected to supervise the gin operations of the new association. The board would consist of three grower representatives from Crosbyton and two each from Wake and McAdoo.

After the presentation, the members adjourned to separate meetings to vote on the proposal. They approved it overwhelmingly. The vote was: Farmers Cooperative Gin, McAdoo — 41 for, 9 against; Wake Cooperative Gin — 44 for, 7 against; Crosbyton Cooperative Gin — 58 for, 1 Against. Of 160 stockholders present, nearly 90 percent stood behind the plans of Dan Davis and the recommendation of their boards. As the formation of a cooperative goes, the resolution and vote were unremarkable. What they portended for the future of those at the meeting and for West Texas agriculture was something else again. No one could possibly have seen where the vote of March 8, 1973, would lead. Some dreamed it. But no one could have seen it.

For nearly two years, American Cotton Growers would have only one division, the Crosbyton Gin Division. There were no others, and with reason. Dan Davis and PCCA planned eventually to build super-gins all across the West Texas Plains, 15 or 20 of them to replace and consolidate the more than one hundred cooperative gins then operating. As new super-gins were built, each would become a division of ACG.

The first meeting of the Crosbyton Gin Division (CGD) Board of Directors took place less than two weeks later. Because the new cooperative was not yet legally established, the seven directors were temporary. They were T. W. Stockton, D. J. Moses and Charlie Wheeler of Crosbyton; Clyde Crausbay and J. K. Edinburg of McAdoo; and Compton Cornelius and G. B. Morris of Wake. Few things are worse in a cooperative than having a community or other group of members feel slighted. This association made every effort to achieve balanced representation. At the first meeting, the officers were elected from each community: Stockton as president, Crausbay as vice-president and Cornelius as secretary-treasurer. The management of the new association came from the PCCA staff, as it would throughout ACG's history. The roles the staff played for PCCA simply were replicated for ACG. Though he never had the title, Dan Davis was the de facto general manager.

On April 24, the temporary board adopted the trademark of ACG, the name AMERICAN COTTON GROWERS in white letters imprinted on the blue background seal. On a motion by G. B. Morris, the directors scheduled a membership meeting for May 4, to elect directors who would replace those named in the articles of incorporation and then go on to serve for the first year. Cornelius moved that the number of permanent directors be seven. The motion carried. The membership meeting also would consider by-laws and a plan of merger between the three ginning communities.

At the May 4 meeting, with 91 members in attendance, J. W. Jackson moved that the boards of directors for ACG and CGD be one and the same. With only the one division in the association, the motion made sense. Seconded by Ross Cash, it carried with only one opposing vote. The process of consolidating the leadership of ACG was under way. T. W. Stockton asked for nominations to fill the board of directors, but Norman Hardy simplified the electoral process. He moved that the temporary board of directors be nominated and elected as the permanent board of directors for ACG and CGD. Ross Cash seconded. The vote was 73 for, 13 against, and five abstaining. The newly elected board members held their first meeting immediately and elected the temporary officers to permanent service. The first board of directors and slate of officers of American Cotton Growers was in place. At the same time, "dirt work" began on the new gin site.

SUPER-GIN

What exactly was the super-gin? In November 1973, soon after the plant went into operation, the PCCA newsletter, *Cotton Cooperative Commentator*, offered a good description. It said the name "super-gin" was only half correct. Super? Unquestionably! A cotton gin? Well, not just that. "Under the [ACG] system, the gin becomes . . . one part of an overall plan that encompasses all functions of moving cotton from the turnrow to the ultimate consumer. By putting together one system with such functions as ginning, compressing and marketing, ACG members fully expect to produce better quality cotton that can be sold for more money at a lower handling cost." The article recognized that "these functions are not profit generating activities instead they result in expenses which [actually] reduce farm income. The real 'profit centers' are the farms themselves.'' This is a significant point. The purpose of the super-gin and related cooperative activities was not to make money but to enhance the ability of the *farm* to make a healthy profit.

A feasibility study conducted by Dr. James E. Haskell, agricultural economist of the USDA, indicated that by eliminating the division of functions, grower-members of ACG could save more than $14 a bale in getting their products to market. Two additional functions, burr pelleting and cotton blending would, according to Haskell, produce total added revenues of $10 per bale, enabling ACG members to save just under $25 a bale with the new system.

''At the time growers become members in ACG,'' the article said, ''they commit all or a portion of their crop to the new coop. This season [1973–74], approximately 45,000 acres were contracted to ACG. Growers can contract all or a portion of their crop and contracts remain in effect from year to year unless the producer decides not to participate.

''Each producer-member of ACG harvests his own crop and stacks it on the turnrow in ricks. When he is ready for it to be picked up, he notifies his area manager [there is a full-time manager for each community], and from this point on the cotton becomes the responsibility of ACG.''

Stockton was quoted. ''This enables the farmer to concentrate his efforts on harvesting. The area managers are not burdened with ginning members' cotton and . . . can concentrate on member relations and loading and hauling operations.''

By late 1973, ACG would operate 150 47-foot containers, each holding 25,000 pounds of seed cotton. These were used to transport the cotton to the gin, where it was handled much the same as it is today. The semi-trailer trucks hauling the containers were equipped with two-way radios to facilitate scheduling with rick-loading and ginning. On arrival at the gin, the containers were loaded onto a hydraulic dump system handling 100 bales per hour. After it was unloaded, the cotton was blended. The blender, considered the most important component tested in the ACG concept, blended the cotton in each container. The process ensured that the 10- to 12-bale load in each container was evenly layered throughout the entire batch. By being blended, bales in each lot contained lint of the same quality. Stockton said, ''Lack of uniformity in High Plains cotton has had a major effect on the costs of ginning, sampling, packaging, classifying, storing, and merchandising. By blending . . . uniformity and some cost problems will be reduced or eliminated.''

Jerry Scarborough was hired on May 22, 1973, to manage the super-gin. Scarborough had worked for Lummus Industries, Inc., the company that sold and installed much of the equipment in the facility. About the gin, he said, ''Blending [allows] gin equipment to run more efficiently. The moisture content in each load is equalized and cotton is fluffed and fed at a controlled rate. The gin is designed to have an average turn-out of 30 bales per hour. Cotton is pressed to universal density enabling it to be shipped anywhere in the world without recompression. Bales are automatically sampled and wrapped in clear plastic shrink film that is not opened after it leaves the gin until it arrives at the textile mill.''

The super-gin would not begin operation until the fall of 1973. In the spring, with ACG newly formed, the gin existed only on paper. In order to build it, the association would need financing. At a May 15 meeting, the officers were empowered by the board to obtain loans on behalf of the association from the Houston Bank for Cooperatives (HBC), ''in an amount not exceeding in the aggregate $2,750,000.00.'' Along with that authorization, they were empowered to ''pledge or mortgage such properties and assets of this corporation as they deem proper, as security for such loans; and to obligate the corporation to purchase Class C stock of said Bank as required by law.'' It was the beginning of a long and sometimes tortuous road that the cooperative and its bank would travel together.

THE COOP BANKS

A special banking system exists for agricultural cooperatives. It is loosely referred to as the ''Farm Credit System,'' though that is a generic, rather than official, title. The Farm Credit Administration [FCA] is a government agency, based in Washington, DC, which oversees the operations of the Farm Credit System. The members of the FCA Board [also called the Farm Credit Board] are appointed by the president and the Congress. The System has several components, one of which is the Banks for Cooperatives (B/C). These banks lend money to cooperative enterprises in their geographic areas. At this time, the B/C system was a network of 13 banks around the United States. There were 12 district Banks for Cooperatives, which were affiliated with the Central Bank for

Cooperatives, headquartered in Denver. The Central Bank handled the portion of large loans that were in excess of the district banks' lending limit. In 1988, the B/C system was restructured. Ten of the district banks and the Central Bank for Cooperatives merged into the National Bank for Cooperatives, called "CoBank." The two district banks in Springfield, Massachusetts, and St. Paul, Minnesota, did not merge; thus, there now are CoBank and two district banks for cooperatives.

The Banks for Cooperatives are owned by the farmers who use them. They lend money at a slight margin. Whatever profit they make goes back to the farmer-owned cooperatives, normally in a ratio of 20 percent in cash and 80 percent in stock in the bank. Representatives of the cooperative organizations that patronize and own the bank sit on its board of directors. Each organization has one vote in the election of directors, without regard to the level at which they patronize the bank.

The bank servicing Texas was the Houston Bank for Cooperatives, the same bank for which Roy Davis worked in the 1930s. In later years, it would move to Austin, the state capital, and its name would be changed to "Texas Bank for Cooperatives," a more accurate title. By 1988, from 40 to 45 percent of the bank's loans were in cotton and cotton-related activities (gins, regional cooperatives, and so forth). The rest were in grain and dairy, with a small amount in sugar and citrus.

In June, 1973, the board heard T. W. Stockton report on a visit he and Dan Davis had paid to the HBC. In Stockton's opinion, the bank was ". . . well pleased with the progress being made by American Cotton Growers, Crosbyton Gin Division." One month later, Don Scott, a loan officer for HBC, reported that the loan requested for construction of the gin had been approved by the bank's loan committee and sent to the Central Bank for Cooperatives and the Farm Credit Administration for final approval.

TRAGEDY AND CHANGE AT PCCA

Soon after the leadership of ACG was in place, that of PCCA changed hands, but not by election. Howard Alford, president of that organization for 14 years, died suddenly. His loss not only created a void but brought to an unexpected and tragic end one long period of strong, unbroken leadership. The election of Alford's successor would prove over the next decade to be one of the most significant decisions by the PCCA Board in its history. It marked the onset of another long period of leadership, this one characterized by creativity and success on the one hand, and controversy on the other.

At the PCCA Board's June 6, 1973 meeting, L. C. Unfred, a producer-member from New Home Cooperative Gin, was elected president. Alford's and Unfred's styles were as different as day from night. Both were strong and determined leaders, each effective in his own way, but that is where any similarity ended.

Alford was blunt, a frank man of few words. He was as unpretentious as he was outspoken. Never one to remain neutral on a given issue, he always made known precisely where he stood. Nonetheless, he was respected by his peers and considered by many of them to be a close personal friend. The words of a resolution passed after his death say it best. "The PCCA Board, is acutely aware of the void created by the loss of our President Howard Alford, for whom we all had the greatest affection and respect. . . . We will not forget the invaluable contribution [he] made to the successes we shared together, nor the courage he gave us in adversity. He was not only our President and business associate but a true and loyal friend to us all. . . ."

Unfred preferred quiet diplomacy in his leadership role. Over the next 13 years in which he would lead PCCA, he was an advocate for fairness. First as president of the cooperative, then as chairman when the title was changed, he was the ultimate arbiter of disputes, the champion of reason, the maker of peace. Whether in the turmoil of controversy or the upheaval of progress, Unfred was the eye in the middle of the storm. He would guide PCCA through development, implementation, and success of the TELCOT electronic marketing system. When the fledgling ACG expanded to take risk and responsibility for an entrepreneurial venture unheard of in cooperative agriculture, he would take the reins of that organization, as well. And when the two cooperatives he led were later at each other's throats, he used tact, wisdom, friendship, and strength of character to try to keep them together. As is usually the case with quiet leaders, Unfred is an easy man to overlook when things are going well. But that is not when organizations need leaders like him. When they are in trouble, when the situation is most critical, when noise is not the answer, they look for an L. C. Unfred. If they are lucky, they have one. During the darkest hours of PCCA and ACG, the cooperatives needed Unfred's leadership, and it was there.

Dreams in Denim

While the Crosbyton gin was being raised, Mr. Roy occupied himself at the Textile Research Center. Jim Parker recalls, "He got interested in some of the work we are doing here. . . . He'd been fishing around with the idea of taking cotton from this area and spinning it on a spinning system that was brand new at that time, open-end spinning. Open-end spinning had been coming along since about 1970, and we got our first machine, I think, in 1973. Mr. Roy was fascinated with it because it eliminates some of the traditional processes. It spins yarn at a much faster rate . . . and the yarn is actually smoother. You've eliminated a process . . . and you've saved on energy, floor space, capital outlay . . . you've saved in every way. . . . He got this idea that we ought to take this new spinning system, use West Texas cotton and make some fabric. That evolved into the denim thing."

The "denim thing" seems to have been on Roy Davis's mind well before he assumed his duties at Texas Tech. A year earlier, he had encouraged and participated in the Lockwood Greene study. As he savored the information reported in that study, his conviction grew. Textile production, he knew, belonged on the Great Plains of Texas. Through the late part of 1972 and the early part of 1973, the elder Davis was instrumental in a second study. Conducted by Lockwood Greene and Booz, Allen & Hamilton, Inc., another consulting giant, this study focused on a particular fabric that might be produced there — denim.

BLUEPRINT FOR DENIM

A July 26, 1973, news release issued by the Farmers Home Administration [FmHA] Texas Office announced, "A joint economic and engineering study made by the U.S. Department of Agriculture and two national consulting organizations has just been completed to determine the feasibility of establishing a cotton textile plant in the Texas High Plains area.

"This study was undertaken at the request of the Texas Industrial Commission, the Natural Fibers and Food Protein Committee of Texas, and local people, particularly Mr. Roy B. Davis, well known leader in the High Plains cotton industry. USDA's Agribusiness Program sponsored and directed the investigations."

Coincidentally, the assistant administrator of FmHA for Business and Industry in Washington, DC and overseer of the program under which the study was conducted, was Denton (Jack) Sprague, a native of Tulia, Texas. Sprague had worked for Roy Davis years earlier at the Plainview creamery operation.

In conducting the study, Booz, Allen & Hamilton Inc. looked at market opportunities for the fabric that such an operation would produce, as well as the availability of labor, water, and other utilities. Lockwood Greene provided engineering data on plant design, equipment needed, operating costs, and expected income for an optimum size plant.

On August 24, nearly a month after the news release was issued, the two consulting firms formally submitted their report. It was everything the visionaries could have hoped for. Not only did it find feasible the establishment of a textile processing plant in the Texas High Plains region, it found that such a plant might also be highly profitable. The area included in the study encompassed the South Plains Planning Region, composed of 15 counties: Bailey, Cochran, Crosby, Dickens, Floyd, Garza, Hale, Hockley, King, Lamb, Lubbock, Lynn, Motley, Terry, and Yoakum.

The first finding of the study, unsurprisingly, was that "the economy of the High Plains region is agricultural, with cotton the dominant crop." At the time, annual production there was averaging more than a million bales, about 35 to 40 percent of the cotton produced in Texas, or 10 percent of total U.S. production. The cotton revenue of the region was $237 million in 1971, one-third of the total income of cotton farmers in the state. Of the cotton produced in the region, 50 to 60 percent was exported. "U.S. textile mills purchase mainly medium and long staple cotton which is necessary for high count yarn and combed cotton goods," the report said. "Foreign mills often produce coarser, heavier goods and find the shorter staple cotton produced in the region suitable for their needs." It projected that cotton "should remain the major area crop for the foreseeable future."

But the dominance of cotton also had its downside. Texas Industrial Commission data showed agricultural employment dropping due to increased mechanization on the farm. At the same time, there was not enough manufacturing activity in the region to pick up the slack. The population of the study area had increased by less than one percent in the 1960s, to 328,000. Lubbock County increased by 15 percent during that period, to 179,000, while the population of the remaining 14 counties declined. The decline in the rural counties was attributed to the trend toward consolidation and mechanization on the farm. In 1961, 36 percent of the cotton in Texas was handpicked; by 1971, only one percent was *not* mechanically harvested.

Growing more cotton was not the answer to the lack of employment opportunities in the outlying counties. Increases in production could have only limited carryover into purchases of services and utilities, because 90 percent or more of the cotton produced was exported from the region. "New industry," the report claimed, "would strengthen the region's economic and employment base." It also would stem the emigration from rural areas to Lubbock.

Nationally, cotton's share of the fiber market had declined. Production was down 30 percent in the previous 10 years, levelling off at around 10 million bales per year. Domestic consumption remained steady at eight to eight-and-one-half million bales. Normally, if production is down and consumption is steady, that means more profit for those still producing. But another trend, an ominous one for all the natural fiber industries, was reported. "During the same [10-year] period, textile mill consumption of fiber had risen 52 percent, from seven to 10.7 billion pounds, but all of this growth had gone to synthetic fiber." The so-called "easy care" products, polyester the major one, were at their peak of popularity. At the time of the report, synthetics accounted for over 60 percent of the total fiber market.

The implications for West Texas were two-fold. Because cotton's market share had declined and then gone flat, cotton grown in the region would be in heated competition with cotton from other regions and even from foreign countries. And because the market share of natural fibers had declined, textiles produced in West Texas would be fighting what seemed to be a national trend toward synthetics. Even if well-manufactured, the fabric made in West Texas would be useless if it was not a fabric consumers were buying. The report said the industry was "highly competitive" and "often troubled." It pointed to 1972 sales at a record high of "over $24 billion, up nearly eight percent from the previous year. However," it added, "earnings of 2.6% of sales have remained at a rate well below those in other manufacturing sectors." Booz, Allen's analysis of the national textile industry seemed to provide every good reason for not going into the business.

But the picture improved as the focus of the report narrowed. "The Texas High Plains region offers locational advantages for a textile plant." Mill owners in Texas would have to meet lower wage rates. Hourly earnings in Texas textile mills averaged $2.27. In the Southeast, earnings ranged from $2.46 to $2.58 per hour. Furthermore, wages for non-durable goods manufacturing in Lubbock and the 15-county area were 20 percent lower than for Texas as a whole. The study reported favorably on availability of labor, water for dyeing and finishing, and raw materials access for plants using locally grown cotton. It estimated the purchase and consumption of Lubbock area cotton locally could save up to $6.80 per bale in selling costs, compression, and transportation.

Narrowing further, the report identified the one fabric both ideally suited for West Texas cotton and still flourishing in the face of competition from synthetics. "Blue denim," it said, "is particularly suited to production in a High Plains textile mill for logistical and market reasons." Denim provided the strongest market for all-cotton fabric. The report called it "one of the fashion phenomena of this time. While production of cotton broadwoven goods has declined by 25% in the last 5 years," it said, "production of denim . . . has increased by 45% . . . ."

The authors expected the demand for denim to last. "Denim consumption has a long history of cyclical up trends and downturns. [But] while denim is currently in the midst of an extended growth period, apparel manufacturers [predict] no consumption downturns in the foreseeable future. . . . Considering both markets and available raw materials, blue denim is a particularly attractive product," the only one suited to West Texas cotton with a solid growth history. Accordingly, the consultants prepared their preliminary engineering and cost analyses "for an indigo-dyed cotton denim manufacturing facility." They concluded such a plant would be highly profitable. "The anticipated return on investment for a denim manufacturing facility," they said, "is approximately 19 percent." This was a heady conclusion for a new firm in a troubled industry.

In another departure from the conventional wisdom, one with which few textile people would have agreed at the time, the report said that the ". . . optimum model plant for the study will utilize the open-end spinning process." Roy and Dan Davis, PCCA, and the TRC must have done a masterful job of convincing the engineers that good quality denim could be manufactured solely through the open-end spinning process. "This process," the study explained, "uses highly automated textile manufacturing equipment and will provide higher unit production and a superior return on investment compared to other processes. The process is particularly suitable to the grade of short staple cotton produced in and close to the study area.

"The model mill," it said, "is designed to produce about 305,000 yards of finished fabric per 120 hours of operation." It predicted actual annual production would be 15.3 million yards of finished denim. "With plant and site costs of $7.3 million, equipment costs of $8.2 million, and working capital needs of $4.6 million, total investment in the facility would approximate $20.2 million.

"Anticipated sales," it said, "are $16.5 million, with profits of $2.4 million after taxes. With depreciation included, cash flow would amount to 19% of investment."

The report predicted that profit at the plant would be more than five times that in the rest of the textile industry as a whole, and three times that in general manufacturing in the United States. These are intoxicating figures, especially in light of the fact that they were based on a production technique, open-end spinning, that was questionable.

Finally, the report claimed, "A new mill has the best prospects of success if affiliated with a major textile [manufacturer]. Texas has a history of textile venture failures. Aside from import competition, the chief causes of failure have been: inexperienced management; undercapitalization; obsolete equipment and technology; and weak marketing."

The antidote to failure might be an arrangement with an apparel manufacturer, which would lead to a secure customer relationship. The report warned that the "sellers' market" of the time was temporary. "In slack demand periods," it said, "competition among producers for markets is intense. During downturns in sales, apparel manufacturers will continue to buy fabrics from . . . regular suppliers, cutting off others from whom they buy incrementally." This would prove to be not only wise but prophetic.

ENTER THE PARTNER

In the early fall of 1973, in San Francisco, an employee of a major apparel manufacturing firm read an article about the study in a trade publication. Werner Pels was manager of product evaluation for Levi Strauss & Co., the venerable blue jeans manufacturer. He was responsible for raw material quality control. His memory of the article is dim, but he recalls that it referred to "some state commission in Texas . . . trying to encourage development [there], particularly from the textile industry. I don't even remember what organization . . . was trying to drum up the trade but they mentioned . . . trying to get a mill to be interested in locating there. It seemed like an interesting thing for us to explore."

At the time, Levi Strauss & Co. (LS&CO) was feeling the effects of a sellers' market. The company could not get enough denim or corduroy to meet market demand. But U.S. textile companies were not prepared to invest in the construction of new denim mills, despite the need for more of their product. Pels contacted some people he knew at USDA for more details about the Texas study. Then he discussed it with his boss, Al Sanguinetti, at that time director of product integrity for LS&CO, who alerted the firm's marketing staff to look into what was happening in West Texas.

Alfred V. Sanguinetti began his career with LS&CO in the summer of 1946 and became a storied figure in a storied firm. He was in charge of all quality control for a company that built its reputation on, more than anything else, the guarantee of quality. He had been in merchandising and in marketing, and was asked, according to Pels, "to take this product integrity position because at the time we had some problems that needed organization . . . and he's a real organization person. After that . . . he went on to become president of TJC [The Jeans Company, the Basic Jeans Division of LS&CO]."

Sanguinetti gave the assignment to John Hunt, then merchandise manager in the Basic Jeans Division, who recalls being told to "Go down and do something about this." Hunt made contact with Roy Davis in the fall of 1973. Soon after, they met at an airport in Dallas. It was the first of many meetings Hunt would have with the West Texas cooperative leader.

Hunt, an Englishman who immigrated to the United States in 1957, joined the LS&CO staff in 1964. Since then, he has seen the firm grow from annual sales of under $100 million to nearly three billion dollars. Today he is director of purchasing for the corporation.

He does not recall exactly how he first heard the name Roy Davis. But he knows now that he never will forget him. His voice softens as he describes him. "Most admirable man I've ever met."

There was no question in John Hunt's mind about the seriousness of Roy Davis in this initial discussion of West Texas denim potential and Levi Strauss & Co.'s needs. If Davis did not feel initially that cooperatives ought to be in the textile business, then he had changed his mind. He may have been stubborn, but he certainly was not unintelligent or unreasonable, especially in the face of opportunity. Perhaps the Booz, Allen-Lockwood Greene study had helped sway him. Many observers feel that his son, Dan, always an outspoken advocate of the move into textiles, had a great deal to do with the change. In any case, having now decided on a course, he pursued it with vigor and determination. If he harbored any doubts, they never showed.

Hunt recalls that Davis had an entourage with him at that first meeting: members of the PCCA Board of Directors; representatives of Booz, Allen and Lockwood Greene; and the Texas Department of Agriculture. Hunt was warmly welcomed. "At the time," he says, "I believe we [LS&CO] were the largest single employer in Texas." He remembers, "Roy Davis was the one who led the discussion." Hunt recalls Davis telling him, in essence, that what PCCA wanted to do was, "take cotton grown in Texas and make it into Levi's jeans in Texas." Hunt's response was that LS&CO was definitely interested, "in having you build a denim mill for us exclusively."

Levi Strauss & Co. had a number of other suppliers. Cone Mills had been manufacturing denim for them for nearly 70 years at the time. LS&CO had encouraged other textile firms to enter the denim market as well, among them Burlington, Swift, Pepperell, Graniteville, and Avondale. But late in 1973, with LS&CO needing all the denim it could get, the company could not convince its older suppliers to build new capacity.

John Hunt's first meeting with Roy Davis in 1973, with the possibility of a market for all the denim a plant on the Texas Plains could manufacture, began to put a foundation under Davis's dream. Over the next couple of years, Hunt would go to Lubbock "at least once a month. Mr. Roy would pick me up at the airport, and he would start driving. And he would talk to me about his ideas," Hunt says, smiling at the memory, "and didn't pay any attention to where he was going. . . ," his words tapering off. Hunt quickly realized the importance of the textile venture to Roy

Davis. He became a frequent visitor to Lubbock and a tireless advocate for the Texas cooperatives. In Hunt's own words, he and LS&CO wanted to "keep the dream going."

Jim McDermott, then general merchandising manager for LS&CO and Hunt's boss, remembers Hunt coming back from his earliest meetings with Roy Davis and PCCA and saying to him, "This thing looks real and you better get your butt down there and find out what's going on, too. These people want to get together with the senior people in Levi Strauss and see if this is for real or if some guy is just playing around with them." McDermott went to several of the meetings with Hunt. He recalls his first. "There was nothing as far as a plant was concerned. It was just an idea. There were just the two of us and there were about eight of them and they took us to lunch." He was skeptical. "I went there not knowing what to expect." The meeting moved him. "I remember that I was impressed and that they were for real . . . there was no question about that. To tell the truth," he adds, "we were a little incredulous because, I guess, it was a rather historic undertaking. I don't know how many years it had been since the last denim mill had been put up in this country, but it was a very adventuresome task they were taking on." After the meeting, McDermott still had doubts, "not about whether they were going to do it . . . [but] whether they were going to make any money doing it."

According to Jim Parker, John Hunt also had doubts, not about the West Texans' intentions but about the process they planned to use. He recalls Hunt saying to a group that included Parker, "Let me tell you something before you get really excited about this thing. I can promise you and guarantee you that Levi Strauss will never, ever use a piece of denim out of open-end yarn. Just not going to do it. If you think that you're going to develop denim that Strauss is going to use out of open-end yarn, forget it!"

Hunt soon realized the nature of West Texas cotton demanded open-end spinning be used. The fiber simply would not permit the use of the traditional ring-spinning technique for the manufacture of denim. "The only way [PCCA] could plan to build a mill," he says, "was to go open-end both ways [in the warp, or vertical yarn, and in the filling, or horizontal yarn]. But that was a process nobody else had used at the time. And the textile industry said that these people would never be able to do it." Hunt looks off in the distance as he thinks back to his original reaction. "There was always the chance that the rest of the textile industry was right, that the technology which they said could not make denim to our [specifications] wouldn't. But [PCCA] had sufficient confidence in Roy Davis and Texas Tech and said, 'We will do it.''' A smile spreads across his face, and he slowly shakes his head. ''It was amazing. Amazing.''

Karin Hakanson is a textile engineer, a graduate of the Finland Institute of Technology. She has been in the United States for over 20 years, most of it in marketing with LS&CO. ''My initial concern was, 'How on earth are they going to be able to set up a mill there with no trained labor and with very little expertise in textile manufacturing?' Textile manufacturing is a highly industrialized field.'' But most of all, she remembers concerns over open-end spinning. ''It was'', she says, ''the number one concern we all had. It was such a revolutionary concept to use it in both the warp and filling. We were sure it just would not provide a strong enough yarn to meet our standards . . . unless you selected a very strong cotton mix.''

Hakanson explains the difference between ring and open-end spinning. Ring-spun yarn has a ''high twist . . . the fibers are quite parallel . . . and it's quite a strong yarn. Open-end, the fibers are not so parallel and . . . you don't have the same kind of twist as you would have with ring-spun. So the yarn is much less strong, and the primary problem we have is tear strength. At the time,'' she recalls, ''our other suppliers had learned to make open-end yarn strong enough for filling, but not warp. So that was a great concern of ours. How would we get the denim that would meet our standards?''

Tom Kasten, who worked for Jim McDermott, had a more blunt assessment of the cooperative's chances. ''Interesting,'' he thought, ''but no way in hell is it going to work.'' Kasten began to change his assessment after meeting PCCA's leadership a few months later. ''It took me about a nanosecond,'' he says, ''to figure out that these were very professional people, very bright, very sharp.'' But he still harbored some doubts. ''These people haven't made a yard of fabric in their history,'' he said. ''What makes them think they can all of a sudden be in the apparel business?''

Jim Parker recalls, too, when the cooperative learned that ''Levi quality'' means having strength, but also the right look and feel. It appeared at first that the open-end process might not only fail to produce fabric that was strong enough for LS&CO, but that lacked the ''right feel.'' Parker explains. ''There is an exit tube where the yarn is spun on open-end machines, and it has an exit navel [with a] hole [in it]. The yarn . . . that's spun is extracted through [the hole] . . . and is wound up into a packet.'' If the surfaces of the navels are smooth, which they were when PCCA began sending LS&CO samples, the yarn will be smooth. LS&CO's response to the samples was not what the cooperative expected.

''They said it was too even, too smooth,'' says Parker. ''They couldn't use it because the fabric was just too good. So we actually took off these little navels, put them in a vise, took a hammer and chisel, and roughed them up. We put them back on the machines, ran some more samples, and sent them to Levi Strauss. They said. 'Hey, that's a lot better.' Today, they make the navels with grooves in them for various yarns. It was only five or six years later, in the late 1970s,'' Parker says, ''that LS&CO was trying to get all its suppliers of denim to make their yarn exactly as it was being made by the West Texans. ''That was five or six years after John Hunt said, 'We'll never use one yard of denim made out of open-end yarn!'''

Although the timing of PCCA's proposed entry into the denim industry was perfect because of the short supply and LS&CO's urgent need, Jim McDermott still was taken aback by the prospect of cotton farmers entering the denim business. The one factor that made him feel they might somehow be able to do it was nothing concrete or really definable. It was the attitude of Roy Davis. McDermott says, ''I guess I wasn't too skeptical after seeing them and . . . the intensity of Roy Davis. He was very serious, very intense,'' about farmers going into textiles. He remembers laughing when he and Hunt got on an airplane to return to San Francisco. ''John, I think that old man is going to be buying us next.'' Still amused by the thought, he says, ''I think if he had lasted long enough, that probably is what his next step would have been. Or he would have decided, 'We don't need Levi Strauss. We'll manufacture jeans ourselves.'''

Who could question the old man's determination? Jim Parker talked with him soon after John Hunt's discouraging remarks about open-end spinning. ''I said, 'Mr. Roy, maybe we ought to hold off and forget about this because I don't think Strauss is interested.''' The response was terse and unmistakable. ''Nope. We're going to do it.'' Parker pursued his argument, ''Well, Strauss is not going to use it. Maybe Haggar won't use it, either . . . or maybe Wrangler or nobody will use the stuff. You've got to have a customer if you're going to go in business,'' he said, pressing harder. ''If they're not going to buy the stuff, you don't need to make it.'' The elder Davis wasn't moved. ''No, we're going to do it.'' Case closed.

The ill-fated Amerasian venture of five years earlier still fresh in their minds, Roy and Dan Davis traveled to Washington, DC with PCCA President L. C. Unfred. They called on Earl Butz, then secretary of agriculture, and Frank Elliott, administrator of the Farmer's Home Administration. They talked with the two officials about their plans for a textile mill and warned them that they could expect ''tremendous pressure'' from the textile industry if the plans went ahead.

Dan Davis says, ''We didn't want to spend a lot of time and money trying to bring it to fruition if they were going to succumb to the same kind of pressure that killed our entry into the textile business in the Far East.'' According to Davis, Butz responded that he was not particularly a fan of cooperatives, but if the project was sound he would not kill it, because it was the first entry by a cooperative into the textile industry in the United States. ''In other words,'' says Davis, ''the project would stand on its own merit.''

On December 17, after a meeting with PCCA, Jim McDermott wrote to Roy Davis at the Textile Research Center, telling him, ''We are anxious and willing to contract for all the proposed mill's first quality heavy weight denim equaling our specifications for the 1st five years of production.'' McDermott doesn't recall how many dollars were involved in the five-year contract, but he says it ''was the biggest commitment [LS&CO] had ever made.'' He credits a number of people at LS&CO with creating the climate there in which the commitment could be made. ''Werner deserves an awful lot of credit for convincing them that open-end spinning could be a reality. [When] my boss, Sanguinetti, . . . got on the bandwagon and said that the open-end spinning yarns could be made acceptable, I think that took care of that.''

Asked how Sanguinetti became convinced that the open-end spinning process would work with West Texas cotton, McDermott says he and Sanguinetti visited the TRC, had long talks with the staffs of TRC and PCCA, and that TRC had sent Sanguinetti samples which he tested. Sanguinetti relied on the sophisticated Levi Strauss & Co. laboratories to test samples, but he had his own way of verifying quality. McDermott says, ''[He did] things like giving them to neighbors' kids'' and seeing for himself how they would wear. With the TRC samples, he became convinced that ''you couldn't see the difference in actual wear. Once that part was done, the rest of it was just getting a bunch of very conservative businessmen, which the owners of our company were, to make a commitment.''

McDermott adds that Sanguinetti, in addition to being a stickler for quality, was an astute businessman. He knew LS&CO needed the denim, and if it tested ''slightly less better,'' it was worth at least a look. One thing in particular about Sanguinetti's role is completely clear. So highly was he regarded in LS&CO, so strong was his influence, that, according to McDermott, ''If he had been against [the PCCA plan], there would have been no sense in even going to the meetings.''

The five-year contract would be one of the three critical elements enabling the mill to be built, the other two being support from the cooperatives and financing from the bank. In reflecting on it today, Pels says, ''I don't know whether [LS&CO's] was a fantastic commitment, considering the [quality] restrictions we put on the output, but we were all very hopeful that it was going to come through. . . .'' Whether it seems in retrospect a ''fantastic commitment,'' it gave added substance to the dream of a denim mill on the High Plains of West Texas.

THE SEARCH FOR MONEY

The agreement with LS&CO would help PCCA get financing, but sources of funding had to be secured. The Farm Credit System, through the Houston Bank for Cooperatives, was the logical source. But unless protected from default, the bank would not even consider a loan of that size. The loan would have to be guaranteed by the federal government. That is where the Booz, Allen–Lockwood Greene study proved its worth. The study's finding that ''New industry would strengthen the region's economic and employment base'' and its glowing prognosis for textile production on the High Plains made PCCA's venture a perfect candidate for support under the Rural Development Act of 1972. That was the law Bob Pogue and George Mahon had labored to develop and move through the Congress. The Farmers Home Administration was empowered under the statute to stand behind business development loans in rural areas for up to 90 percent of their face value. It did not take long for the Plains Cotton Cooperative Association to submit a project proposal to HBC and FmHA.

Over at American Cotton Growers, attention was focused on the super-gin, which, at the height of the ginning season, still was not operating at full capacity. At the January, 1974, board meeting, Dan Davis reported that the compresses were 60 to 90 days behind in their shipping orders. Cotton was backing up, and it would affect the money position of ACG if the ricked cotton were not ginned soon. ACG would have to gin its cotton somewhere other than the super-gin. Within days of the meeting, the board voted to allow ricked seed cotton to be ginned at the Mt. Blanco Gin and the Dougherty Gin in ACG's name. The irony of this cannot be appreciated fully until one realizes how much time and effort the cooperative later invested in enforcing its contract with members, one that prohibited them from ginning their cotton at non-pool gins.

Dan Davis told the ACG Board in April that it might be 1975 before another super-gin would be planned and another division of ACG formed. Between marketing cotton, bringing the super-gin on line, planning a denim mill, and other activities, the PCCA staff, which also served ACG, was on overload. He explained the proposed textile mill "that might be built by the five regional cooperatives," with PCCA in the lead. He told the board that growers might realize $50 a bale more over the market value of their lint through the sale of denim that the plant would manufacture. The ACG directors listened attentively, but at this point the prospect of PCCA building a denim mill was of only marginal interest to them.

On March 22, 1974, J. Lynn Futch, head of the FmHA Texas State Office in Temple, wrote the FmHA County Supervisor in Lubbock. He said the Booz, Allen–Lockwood Greene study " . . . indicated the [PCCA] project merits further consideration . . . and that . . . the applicant should be requested to complete the formal application, provided it is determined that all essential elements for feasibility are present." One of the requisites was that "a determination of feasibility by the proposed lender should be made prior to further processing of the guaranteed loan request." The HBC would have to consider the loan before FmHA would consider the loan guarantee.

In May, Bill Blackledge, as legal counsel to PCCA, wrote Ormel I. (Jack) Boyd of the FmHA Texas Office. He enclosed a copy of Jim McDermott's December 17 letter. Blackledge told Boyd that PCCA expected to "receive a firm commitment while reserving the right to sell a percentage of the plant's output to other customers in order to build up a market."

Boyd had been reviewing PCCA's proposal, dated May 20, 1974. It requested $28,242,000 to be used for "production of cotton denim fabric beginning with raw cotton." He focused on all the pertinent points. The cost of the building was estimated at $11,157,300, to be repaid over 25 years; equipment at $14,759,976, to be repaid over 12 years; and working capital at $2,175,000, to be repaid over seven years. Land acquisition was estimated at $150,000. The total annual payment to the bank would be $1,994,981. PCCA offered as collateral its net worth of $11,822,214, virtually all its assets, including the compresses at Sweetwater and Altus. The number of jobs to be created was estimated at 380: 300 in the labor category and the remaining 80 divided between 50 skilled, 15 semi-skilled and 15 management and administration personnel.

Boyd was developing a "profile" of the project his agency would use in reaching its decision. FmHA had not worked with PCCA before, and he had to rely on opinions of third parties. In his analysis, he described PCCA's present management as "demonstrated," citing an opinion he received from HBC. The bank told him the cooperative's "management has demonstrated a high degree of competence, reliability, integrity, and efficiency in conducting its business affairs." He acknowledged that the textile staff, a very critical element, would have to be secured. Boyd's profile reported the proposed facility, located in an area of over two million bales of annual cotton production, would have no problem securing raw materials. This assumed the plant would have access to all the cotton it needed. It was an assumption Jack Boyd could not then have known would prove to be less than sound.

The marketing system was described as, "Initially a five-year contract with Levi Strauss [that] will provide for delivery of major portion of cotton denim produced to that firm. Price to be based on a weighted average of price paid to other major Levi Strauss suppliers." He noted that "a firm contract [would] be a condition of [FmHA's] commitment." Prospects for repayment were called "good, as assessed by the Houston Bank for Cooperatives and projected by Lockwood Greene Engineers." Again, the study helped. Boyd projected the break-even point "between [the] 23rd and 24th month of operation when finished yards of denim reach between 531,743 and 754,732 per month and estimated profit reaches $390,226 per month. This monthly level of profit," he wrote, "is projected to be adequate to pay new debt." Boyd recommended his office request authority to commit the guarantee conditionally, subject to site approval and concurrence by HBC and FmHA in the sales contract with Levi Strauss & Co.

From his office in Washington, Congressman George Mahon closely monitored the status of PCCA's application. Levi Strauss & Co., too, kept a watchful eye on the situation. Mahon's papers show Jim McDermott called him on June 6, 1974, to inquire about the status of the loan guarantee.

On that same day, and possibly through the urging of George Mahon, Jackson E. Simpson, director of the Agri-business Program, Agricultural Research Service, USDA, recommended approval of the textile mill proposal in a letter to John Swinnea, Jr., director of the Business and Industrial Loan Division of FmHA. Simpson's agency had commissioned the Booz, Allen–Lockwood Greene study, but Swinnea's would have to approve the loan guarantee. In the letter, Simpson described the objectives of the study commissioned by his agency. He emphasized the investment in the study made by USDA, the department for which he and Swinnea both worked, as though to ensure the results would be put to use.

"The current project," he said, "is for a $28 million plant to produce denim fabric that will be sold to such large consumer manufacturers as Levi Strauss. This objective is directly in accord with the Agri-business findings and recommendations." He concluded, "Under proper management, we are . . . of the opinion that this type of project should be a very successful answer to the needs of both the people and the utilization of this large-supply commodity [cotton]. We recommend favorable consideration to this project proposal."

On June 24, Simpson sent a copy of the letter to William Erwin, assistant secretary for rural development and his supervisor, adding "In view that ARS Agri-business Program planned and directed this extensive feasibility investigation, we would think that consummation of this proposal could be rapidly accomplished."

THE "CATCH"

By nearly all accounts, PCCA was well on its way to building a denim mill on the Texas Plains. But there was a catch. According to George Mahon's papers, USDA expressed concern about a complaint against PCCA filed by the Commodity Exchange Authority (CEA), the predecessor of today's Commodity Futures Trading Commission. The issue was raised during a June 11 meeting between Mahon, his staff, and USDA officials. Dr. James E. Bostic, a deputy assistant secretry of USDA, told Mahon that

FmHA would not proceed with the loan application processing until the complaint was resolved.

Dan Davis explains the CEA complaint. "During the fifties and sixties the price of U.S. cotton was very tightly controlled by the government. You had the loan as a 'floor' and the government's resale price as a ceiling. During those days, most every year the government had very heavy surplus supplies of cotton on hand and they sold that cotton within about a cent a pound typically of the loan level. So the price could only fluctuate in a very narrow range, and that remained true through almost all the period of the fifties and sixties. There were some aberrations, but not many.

"In early 1971, the government sold out of cotton, and a scare developed that there wouldn't be enough to go around. The price became very volatile. PCCA always had a policy of trying to sell about one-third of its anticipated production prior to harvest, about one-third during the harvest, and about one-third after the harvest. We had about 225,000 bales sold going into the active harvest period, and to hedge that we bought an equivalent volume of cotton on the [exchange] board.

"It's very difficult to hedge short-staple, stripper cotton because the futures contract is on long-staple, picker-type cotton. The basis [the difference between the New York cotton futures price and the spot market price] can kill you, which is what was happening to us. We had a smaller crop that year, and the price of short staple cotton had risen rapidly and very high, but the futures had not appreciated as you would expect when you're trying to run a hedged operation.

"At any rate, one of the major merchants tendered 80 or 90,000 bales of cotton on the New York board, which had the tendency to depress New York [prices] while the country prices were still going up. We were really getting crucified, and so we arranged a $10 million special line of credit at the [HBC] to take possession of that cotton as it was tendered. Rather than liquidating our futures position, we took delivery, which stabilized the market and caused it to start up. And then we decertificated that cotton and sold it and came out very well on it. But by that time, it was apparent we had the dominant long position in the market, and the competitors of the cooperative had never had a cooperative take an active part in the futures market. At least some of them were very determined they were going to teach us a lesson.

"We really didn't have any experience in handling a major futures position; sure did a lot of things stupid when we should have done better. At any rate, there were 45,000 bales or so tendered in March and, I think, 35,000 in May, all of which we stocked, decertificated and sold on the advice of our attorneys.

"But some of the other merchants filed a complaint with the CEA that we were manipulating the price of cotton and causing it to be artificially high. [CEA] sent people out to talk to us and investigate us. Finally, we had to go to New York and meet with the New York Cotton Exchange officials. Between us and the Cotton Exchange, we worked out a settlement price which they thought was fair and we thought was reasonable, and we settled the May contract at that level.

"We were trying to get out of the market, liquidate our position. But then a lot of cotton was tendered in July and we stopped about half of it, I guess. Anyway, about a year later, the CEA brought a complaint against us alleging that we improperly got the price of cotton too high in May through July, which we contended was not the case. I thought our attorneys did a masterful job of explaining the situation, but we had a law judge that was very negative, to say the least. We settled the case without admitting wrongdoing, but we did accept a penalty to refrain from trading in the futures market for a year."

When the subject of the textile mill arose, the case had not yet been settled, and USDA gave the cooperative something to think about. Recalls Davis, "They said if we lost the case, then we might be the target of civil litigation and there was no way to really quantify the amount of damages that might occur if we lost. And so they would not let the [textile mill] proceed in Plains' name. After thinking about it a long time, I decided, 'Well, we'll try to switch the thing over and have American Cotton Growers build a mill.' By that time, we had negotiated the contract with Levi Strauss, which was really the keystone to the whole deal . . . having a market for the production of the plant for the first five years. We had done all the engineering studies and were a long way down the road."

T. W. Stockton remembers it a little differently. He maintains that he "argued with Dan on several occasions that the proper place for the denim mill to be built was in ACG . . . that we were already risking about everything we had in building the gin plant, which cost $3.25 million, a figure unheard of for a gin at the time." He says that after the CEA complaint became an issue, Dan Davis called him one morning "about 10 o'clock and asked me if I was still interested. I told him I definitely was."

The ACG Alternative

The minutes of the June 18, 1974, meeting of the ACG Board include a terse notation that "a discussion was held concerning the possibility of American Cotton Growers building a textile mill." This was only seven days after the meeting in Washington in which Dr. James Bostic expressed USDA's concern over the suit.

Because PCCA could not obtain federal financial backing for the construction of a mill, ACG seemed a logical alternative. The three communities had gone together to form the super-gin and a marketing pool. They already had received USDA Commodity Credit Corporation clearance to operate as a pool marketing cooperative and utilize the Form G loan program for cotton, enabling them to put cotton into the government loan program and market it over time on behalf of their members. After 15 months, they were established as a cooperative entity.

Later that month, the ACG Board heard a proposal from the West Texas Regional Cotton Cooperative's Textile Committee, a group established by the regionals to follow up on the Booz, Allen-Lockwood Greene recommendations. It recommended ACG request a loan guarantee under the Rural Development Act for the organization and construction of a denim plant. It also recommended that several of the regional cooperatives, including PCCA, Plains Cooperative Oil Mill, and Farmers Cooperative Compress each acquire building bonds of ACG in the amount of $1 million to provide sufficient equity to permit ACG to build the mill.

After a long discussion of the proposals, Clyde Crausbay moved that ACG apply for a loan guarantee of $28,242,276 to build the textile mill and that T. W. Stockton be empowered to act on behalf of the association in meetings with HBC and FmHA. The motion was seconded by Charlie Wheeler and carried unanimously. The meeting then adjourned. Though it hadn't the slightest idea how to get there, American Cotton Growers (three towns and one gin) appeared headed into the textile business.

Planning for the mill went full speed ahead. The PCCA staff, particularly Emerson Tucker and Dan Davis, repeatedly met with the staff of Lockwood Greene, working from the specifications of the textile study to design the new mill, making changes where needed. They identified the equipment around which the mill would be built and its manufacturers.

Lockwood Greene's notes taken during and after the meetings cite "owner instructions" to design production on the basis of seven-day-per-week operation. They also specified the building design "will not include provisions for future expansion," a restriction that was propitiously ignored.

The proposal that ACG would build the mill must have reached the Department of Agriculture almost immediately. On June 29, Frank B. Elliott, Adminstrator of FmHA, sent a memo regarding the project to Lynn Futch, his state director in Texas. "The project," he said, ". . . appears . . . feasible and has sufficient priority for your further consideration and processing."

Elliott, a retired Air Force officer known as "General Elliott," said a contract of guarantee could be executed, subject to certain conditions. He set down eight conditions to be demanded by Futch. Some would become major points of contention between his agency, the Houston Bank for Cooperatives, and the cotton growers.

1. The FmHA reserves the right to concur in the management personnel of the American Cotton Growers.
2. The FmHA approves the project site location.
3. The members of the "Cooperative Textile Committee" i.e., Plains Cooperative Oil Mill, Farmers Cooperative Compress, Plainview Cooperative Compress, American Cotton Producers [*sic*], Plains Cotton Cooperative Association and the Growers Seed Association, also co-sign notes for this loan.
4. The borrower and lender have agreed to provide the FmHA County Supervisor at least every 90 days a consolidated and divisional balance sheet, profit and loss, and source of working capital statements, certified by an officer of the corporation and such other information and documents as FmHA may from time to time require.
5. The required character clearance has been obtained.
6. Your conditional commitment should contain a condition that the Department of Labor certification must be received prior to issuance of the final FmHA guaranty.

7. Members of the "Cooperative Textile Committee" will purchase Building Bonds in the amount of $3 million which will be subordinated in payment and security to the lender and FmHA.
8. American Cotton Growers obtain a purchase order commitment from Levi Strauss and Company for a five (5) year period for the entire production of the applicant's plant.

He acknowledged the Texas FmHA office had obtained a favorable Dun and Bradstreet credit report on ACG, and "We have requested the finance office to obligate guarantee authority for this loan. . . ."

On July 3, T. W. Stockton wired HBC president Murrell Rogers to inform him of the message from Elliott. He wrote, "Farmers Home Administration advises that the application of American Cotton Growers for a loan guarantee to construct and operate a denim manufacturing plant has been conditionally approved." He then recounted the eight conditions FmHA had established for providing the loan guarantee. "This matter will be considered today by the High Plains Textile Committee," Stockton added. "We would appreciate knowing the attitude of the Houston Bank for Cooperatives concerning the proposed conditions to the FmHA approval."

Everything began to unravel at once.

THE FINANCING FALTERS

Rogers responded to Stockton that same day. "After careful consideration of these conditions," he said, "this bank's executive committee finds a serious objection to our making this loan, if three of these conditions are met."

1. "That FmHA have the right of concurrence in the selection of management." It is this bank's practice to maintain this prerogative as normal to our service of any loan even where we have another bank's participation or should have a guarantee.
2. "That the character references of management employees and of each new director be subject to check by FmHA." As to management employees, this is a prerogative of management and we have a degree of influence over such management as indicated above. As to a lender's right to check a borrower's directors, we have practiced this as a routine matter but without any prerogatives as to control over any director's selection or his service on the cooperative's board.
3. "That ACG secure a purchase order agreement from Levi-Strauss covering all of the out-put of the proposed denim plant during the first five years of operation." As the lending bank, we would, of course, require

some assurance of a market for the output of the proposed denim plant but in our opinion it would have to be more flexible than this.

Three of the eight conditions set down by FmHA were unacceptable to the bank. He continued: "Our interpretation of the intent of Congress in its use of experienced lending institutions for such guaranteed loans is that our proven expertise in the making and servicing of loans is to be considered by [FmHA]. As the lead bank, it would be our responsibility to approve and service the loan with the same judgement we impose upon other loans for which there is no government guarantee.

"From the above conditions that are imposed upon this approval, we sense a plan for the operation of the equivalent of two loan committees. If this be the case, our experience in this area demands that we withdraw our offer to finance the manufacturing plant for American Cotton Growers as has been submitted."

The bank made clear that it would not compete with FmHA in setting standards and evaluating performance on the loan. ACG was caught in the middle of a bureaucratic turf war. And there was more bad news.

On the same day, L. C. Unfred, in his capacity as chairman of the High Plains Textile Committee of PCCA, wrote Stockton. Unfred, who later would serve simultaneously as chairman of both PCCA and ACG, told Stockton the High Plains Textile Committee reviewed the Elliott letter that afternoon. "After discussing the conditions of the proposed guarantee," Unfred said, "the committee concluded that we cannot agree to the condition that all regional cooperatives represented on the Committee co-sign the note.

"The total equity capital of these regional cooperatives," he explained, "is $46,500,000, made up as follows: $20.9 million — Plains Cooperative Oil Mill, $7 million — Farmers Cooperative Compress, $2 million — Plainview Cooperative Compress, $15.6 million — Plains Cotton Cooperative Association, $ 1.0 million — Growers Seed Association.

"The members of ACG are members of the High Plains regional cotton cooperatives," Unfred recognized. "However, ACG members own only a portion of the outstanding regional stock. It is not reasonable for all assets of all members of these Associations to be used to guarantee repayment of the proposed loan, since the profits of the mill will go to ACG members alone."

But had not PCCA encourage ACG to apply for the federal loan when it could not? Unfred acknowledged that it had. "You will recall that the PCCA application for an FmHA loan guarantee was rejected. Rather than see the project die, it seemed desirable for an entirely different group to build the mill.'' But that was before PCCA and the other regional cooperatives realized how exposed the government's conditions would leave them. ''The High Plains regionals,'' Unfred continued, ''agreed to invest $3 million in subordinated building bonds in your venture. This seemed a reasonable commitment of capital in relationship to the share of regional patronage and equity American Cotton Growers provides.''

''We must consider the fact,'' Unfred said, ''that obligation of all assets of all regionals to the denim mill project would make it difficult for each cooperative to secure the line of credit necessary to carry out its particular function.

''We regret that we are unable to comply with this condition you have reported to us.'' The regional cooperatives would help, but they would not pledge all their assets, even if their refusal to do so killed the deal. In one day, ACG was notified that its loan application had been approved with eight conditions, but half of them were unacceptable to two of the parties critical to the transaction.

Two days later, on July 5, Stockton penned what seemed to be an obituary for the textile mill. In a letter to Ralph Griffitts of the FmHA Lubbock Office, he reported, ''The responses received from the Houston Bank for Cooperatives and from the High Plains Textile Committee are enclosed.

''Based on these comments, it appears that there is no possibility that the requirements for an FmHA loan guarantee can be met.

''The conditions imposed would interfere with the loan management rights of the Houston Bank for Cooperatives on which cooperatives in Texas have relied for many years for advice and assistance in connection with their operations.

''The requirement that all High Plains cotton cooperatives co-sign the note is not fair to those members of the regionals who would not participate in the project.

''We regret that we are unable to proceed under the conditions proposed by [FmHA].'' If an alternative approach were not found, the dream of a cooperative textile mill on the Texas Plains would be, at best, delayed — at worst, destroyed.

George Mahon, watching events unfold from Capitol Hill, was not about to let this happen. On July 30, 1974, a conference was held in General Elliott's office in Washington. What is not mentioned in official records is that Congressman George Mahon attended the meeting. Though this may seem a fine or irrelevant point to some, it surely is not. Members of Congress normally hold meetings in their own offices. For a member to go to an agency for a meeting is an unusual and dramatic display of interest. When the member is chairman of the House Appropriations Committee, with the fate of every agency's budget in his hands, the impact is overwhelming. By his half-mile journey up Independence Avenue in Washington, DC, Mahon sent a signal to USDA that was unmistakable.

Mahon also had to endure a certain amount of discontent from his southeastern colleagues. His records reflect a February, 1975, call from Congressman Phil Landrum, a Georgia Democrat, who was chairman of the House Ways and Means Committee. Landrum was concerned that the proposed mill would take business away from his home state and the Carolinas. The discussion was between two of the three most powerful members of Congress at the time. Mahon managed to assuage his colleague's concerns.

Jack Boyd was at the July 30, meeting at FmHA. He says Elliott, Murrell Rogers, and a host of federal officials were there, along with T. W. Stockton and Bill Blackledge of ACG and PCCA. Boyd recalls the visit by the venerable Mahon got the close attention of everyone in the room. More important, it provoked a reexamination of the conditions set on the ACG loan.

Two days later, a memorandum from Elliott to William Erwin portrayed the problem as a "communication" failure between ACG and the Houston Bank for Cooperatives. "All . . . questions [the meeting participants] raised but one resulted from inadequate communication between the loan applicant and the lender, that is, the Bank for Cooperatives." This is a difficult accounting of events to accept, because the information conveyed to the Houston Bank for Cooperatives by T. W. Stockton matched, virtually word-for-word, that in FmHA's letter of conditional approval.

But never mind. If it were a face-saving device for FmHA, so be it. Three of the four conditions (all those to which HBC objected) were erased.

The one that still existed was major. Elliott's memorandum continued: "Both lender and applicant representatives explained that they could not meet our requirements that all the other cooperatives in the plains area sign the note with the American Cotton Growers. To do so would result in each of those cooperatives becoming fully liable for the full amount of the $28 million loan." He concluded, "They have a legitimate point."

Jack Boyd agreed. "It really was rather inappropriate for the Department [USDA] and the agency [FmHA] to have asked all those other regionals to join in that, because it was clearly going beyond . . . the intent of the law so far as collateral." He says many others shared that point of view and cites L. D. Smith, a long-time FmHA official. Smith said at the time, "Hell, we're asking that they mortgage all of West Texas to get the financing for this arrangement."

The impetus now was with ACG. The cooperative had to develop an alternate proposal that would satisfy FmHA's last remaining condition, the one aimed at having the farmers who were borrowing the money bear as much of the risk as possible.

On August 7, Bill Blackledge proposed such a plan in a letter to Ralph Griffitts, "as a result of the conference held on July 30, 1974 in Mr. Elliott's office." The plan was based on what later became the most significant issue underlying the operation of American Cotton Growers for much of its future, the "100 percent pool concept."

Blackledge said "The High Plains Textile Committee recommended that the cooperative gins in the West Texas area be contacted to determine if arrangements could be made for a group of gins to participate in a 'seasonal pool' marketing program similar to that of American Cotton Growers [for the super-gin], and [thereby] provide a portion of the required equity investment for the textile mill."

He explained how the system would work. "[HBC] indicated the project might be financed by [them] in connection with seasonal pool commitments of growers' crops with an equity base equal to 20% of the loan." If the regionals were not going to co-sign for the loan, HBC wanted the farmers to put enough of their money at risk to cover 20 percent of it. That later was lowered to 15 percent, which seemed more logical in that 90 percent of the bank's exposure would be covered by the FmHA loan guarantee.

Blackledge estimated the loan would be about $30 million because of "increases in machinery costs." The total equity required, then, would be about $6 million. It would have to be raised from a sufficient number of producers with acreage enough to guarantee 250,000 bales of cotton per year. Why 250,000? It was estimated that the mill would require 65,000 bales of cotton per year to operate efficiently and profitably. It also was estimated that only one in four bales would be of suitable quality for use in the mill. The rest would have to be marketed elsewhere. To form this seasonal pool, more gins would have to be brought in as participants. American Cotton Growers, which consisted only of the Crosbyton operation, had a net worth of $2.6 million to put up. Blackledge estimated that another $2 million would come from "seasonal pool gins other than ACG." This would leave $1.5 million to be raised from other sources, which meant the regionals.

Blackledge's letter to Griffitts continued. "It appears that a number of cooperative gins are willing to contribute funds to be invested in the project in an amount equal to $10 per bale for their average ginnings during the preceding 5-year period." Thus, a gin averaging 10,000 bales a year would sign a $100,000 note to guarantee their portion of the $6 million. How Blackledge knew that "a number of cooperative gins" were willing to do this is not a matter of record. Subsequent events show he may have been either overly optimistic or trying to put the best face on an uncertain situation.

He told Griffitts how the arrangement would be structured. "Patrons of gins desiring to participate in seasonal pool marketing operations will enter into contracts with the cooperatives to deliver *all* their cotton to *participating* gins [emphasis the author's] for ginning, and to deliver their lint cotton [the pure cotton left after ginning] to a seasonal marketing pool from which those qualities needed by the textile mill will be selected to be converted into denim.

"The remainder of the cotton delivered to the seasonal pool will be marketed through the existing cooperative system under the direction and control of a seasonal marketing [pool] committee consisting of one producer from each participating gin association. Participating gins will acquire building bonds for the project in an amount equal to $10 per bale for the volume of cotton committed to the seasonal pool.

"The remaining equity requirements will be provided by one or more of the regional cotton cooperatives involved in the program." Thus, the regionals would be equity participants, but only to the extent that they would make up the difference between what the bank required and what the participating gins could provide.

The meaning of the words in this proposal is critical. Any farmer participating in the seasonal pool would be required to gin *all* of his or her cotton from contracted acres at a pool gin. There would be no choice, no option. And when a gin signed into the program, when it became a pool gin, it could accept cotton only from farmers enrolled in the program. It was what became known as "the 100 percent concept." As a producer, you committed a certain amount of acreage to the pool and you delivered every bit of cotton on that acreage to a pool gin. As a pool gin, you took only pool cotton. You either were in the pool or you weren't. Without 100 percent participation, there could be no system. Only with enough acreage to guarantee 250,000 bales of cotton per year could the textile mill be operated. Only with a sufficient commitment of growers' crops to the equity base could the financing be obtained.

Blackledge, who along with others on the joint PCCA–ACG staff had invested tremendous effort into finding a creative solution to the equity dilemma, proposed the methods of distributing earnings, establishing policies, and controlling operations. "Earnings from denim manufacture [known as 'mill outturns'] and from the sale of cotton not used by the mill [known as 'lint outturns'] would be distributed pro-rata [in the form of 'progress payments'] to all growers delivering cotton to the seasonal pool.

"A portion of the earnings would be retained for debt repayment [to HBC] and capital accumulation [for establishing a healthy equity-to-debt ratio and for operations], with the bank approving all cash distributions to be made to participating growers." The requirement for bank approval of cash distributions was sound policy and necessary for ACG to get its loan. But later it, too, would become a point of contention.

"As capital accumulations permit," Blackledge continued, "the [ACG Board] with the bank's approval, would retire the outstanding building bonds. Participating growers then will own all equity interests in the textile mill." As ACG tried to repay the bank and establish a strong capital base, it also would try to pay back its members through progress payments to growers and in liquidation of building bonds purchased by the cooperatives. It was a heavy burden to be taking on.

In concluding, Blackledge said, "The participating gins, whose patrons become members of ACG, would be grouped by territorial districts from which directors would be elected to serve as the ACG Board of Directors. . . . Each gin will be entitled to be represented on the seasonal pool marketing committee which will [establish and approve] sales policies for the marketing of seasonal pool cotton. The [ACG Board] will be responsible for operation of the proposed textile mill and the establishment of policies for ACG business operations."

That was the essence of the alternative proposal. Condition 3 would be eliminated to avoid pledging the assets of growers who did not want to participate. Instead, more would be pledged by those growers who did want to participate. ACG also suggested a softening of Condition 8, in which FmHA required a five-year purchase order commitment from LS&CO for the total production of the mill. It proposed "purchase order commitments from Levi Strauss or other potential customers for sufficient output of the plant for a five-year period to assure successful operations." The cooperative seemed aware it would take time for "total production" of the mill to meet LS&CO standards, if in fact it ever could. The proposal came from ACG, but it had the "fingerprints" of Murrell Rogers and the staff of the Houston Bank for Cooperatives all over it. The 250,000-bale pool would ensure that the mill had the raw material it needed to operate. The $10 per bale investment ensured that the project would not be dangerously leveraged.

If the bank had concerns over the size of the cotton pool, they were justified. The proposal bearing Bill Blackledge's signature assumed 50,000 bales coming from Crosbyton alone, albeit in "normal seasons." Only three weeks after ACG submitted its proposal, it was reported that, ". . . certified acres planted by patrons of the Crosbyton Gin Division were 43,067 acres. Lost acres due to hail, dry weather, etc., was estimated at 4,678 acres. It was estimated that on the 38,389 acres remaining, approximately *23,818* 500-pound bales of cotton would be produced." [Emphasis the author's] The proposal assumed a 50,000 bale crop from Crosbyton in a normal season. But the bank had to wonder. How many seasons in agriculture are "normal?" And what happens to the textile pool when production is less than half of normal, as in this case? It seemed clear to the bank that in this, as in similar enterprises, more was better in terms of participants. With more farmers and gins in the venture, there would be more cotton from which to select. And with more financial participation in the venture, exposure to failure and financial catastrophe would be lowered for everyone, including the bank.

FILLING THE POOL

Murrell Rogers must have known how difficult the "100 percent concept" would be to sell the farmers. Everyone must have known. Texans are "rampant individualists." West Texans, if anything, are worse. And farmers never have been known to acquiesce peacefully and quietly to any policy or rule they didn't think was wise or fair. Convincing enough gins to convince their members to join the pool would be difficult. Telling West Texas farmers that they had to join the pool or they could no longer gin their cotton where they had been ginning it for years would not go over well.

Dan Martin remembers a meeting in Dan Davis's office. "C. L. [Boggs] and I were in there, and Murrell Rogers . . . was there. Dan and Murrell were arguing about the 100 percent thing. "Dan [Davis] said, 'I don't know if we can get enough gins to agree to get 100 percent of their membership to do this. This is a radical change from what we've ever even talked about in West Texas.' In arguing back and forth," Martin says, "I think Murrell lost his cool. He got up and shook his finger in Dan's face and said, 'If you want that damned denim mill, this is the way it's going to be. Otherwise, forget it! Murrell just turned and walked out the door. We just sat there for a few minutes, which seemed more like 10 minutes. Dan's face was like the mercury in a thermometer . . . the tone of his voice never changed, but when he was getting upset you could watch his face . . . it would just get red. Finally, he looked up and said, 'Fellows, how many gins do you think we can sign up?'

"It seemed we sat there for another five minutes," adds Martin, "and I said, 'Well, I can think of one.' C. L. looked at me and said, 'Who's that?' The one I thought might go along was the gin at New Home because L. C. Unfred was president of their board there and he was our [PCCA] president here.

"Dan Davis laughed and said, 'You know, fellows, we'll never know until we try. Let's give it a try.'"

If the textile mill was to be built, Dan Davis, C. L. Boggs, Dan Martin and others from the staff of PCCA had less than two months to form the seasonal pool. All they needed was the approval of the PCCA Board to go ahead, a great deal of patience and endurance, and no small amount of luck.

On September 11, 1974, Murrell Rogers wrote to Lynn Futch of FmHA in support of the ACG alternative proposal, which he had played a major role in developing. He explained the 250,000-bale pool was, ". . . not only the key to the needed financial stability for the cotton mill project but it is a tool for building stronger cotton cooperative operations of growing, handling, ginning, marketing, and compressing as well as milling of the cotton seed and merchandising the resulting products." His estimate of the time it would take for ACG to put together its 250,000-bale pool, "through many meetings with cooperative farmers in the Lubbock area, is the end of the calendar year." In fact, the conditional FmHA loan guarantee commitment would expire on November 1. ACG had about six weeks to bring in the gins, not 14 as Rogers thought.

In mid-September, the PCCA Board gave the staff permission to start meeting with gins and trying to recruit them for the ACG pool. Because ACG officially could not take responsibility for the mill until its own membership approved that action, the PCCA staff would do its recruiting under the aegis of PCCA. It would be as though PCCA still was the cooperative building the mill. It was a sensitive situation. Making matters worse, the staff had no idea how many gins they were trying to bring in, only that there had to be enough to provide the total annual average of 250,000 bales. One gin might have an annual average of 5000 bales, another 20,000 bales. However many gins it meant bringing in, the goal was crystal clear.

Dan Martin remembers, "The [PCCA] Board's attitude seemed to be, 'If you can get enough gins to agree to it . . . if the bank will agree . . . if the FmHA loan people will agree, okay. Go ahead.' Most of the thinking at that time seemed to be, 'Aw, you're wasting your time, but it doesn't matter. It's your time, you can do whatever you want with it.' A few thought it was well worth looking into, but they seemed to be in the minority."

Martin recalls how the first gin was brought into the pool. "It was immediately after the [PCCA] Board meeting where we were authorized to begin work on this. Jay Cannon, chairman of the Hale Center Coop Gin's Board of Directors, came up to Dan [Davis] and C. L. [Boggs] and me and asked, 'Could y'all meet with our board tonight to explain this?' Dan and C. L. both had commitments that evening. Dan turned to me and asked, 'Would you mind taking Dad [Roy Davis] and going up there and meeting with that Hale Center Coop Board and explaining to them what we're trying to do?' I said, 'I'd be happy to. Hell, I can tell them all I know in about three minutes.'"

He describes the meeting. "Roy and I went and met with the board that night . . . I mean the night of the very day that the PCCA Board voted to let us pursue this. Well, Roy started out. He said to me, 'I know all of those guys. Jay Cannon's an old friend of mine, Dan. I'll talk to them and try to explain it to them.' He got up in front of the group and went through what it was we were talking about doing."

Martin is touched by the memory. "They went to asking him some questions and he really started getting confused, bless his heart. Finally, he said, 'Dan, I don't know what the hell I'm talking about. Why don't you explain it to them?' Of course, I was just as confused as he was, probably more." Being confused didn't stop Martin from making the effort. "They had a blackboard on the wall in that gin office where we were meeting, and I thought, 'If I can start drawing and explaining to them what we're trying to do, we can all follow it a little better, including me.' I got up there and drew circles and explained the 100 percent concept from preliminary figures that we had put together, the dollar earnings they could be looking for out of the denim mill, and how it would enhance the value of their cotton." He admits his figures were less than scientifically derived. "We bounced a ball off this wall and another one off that wall. It was just creative accounting."

Yet the session was not without value. "I explained," says Martin, "how the denim market doesn't fluctuate like the cotton market. It's not as drastic. It's more stable. I said that, during times when the cotton market drops all the way to the bottom, the denim market will decline but not nearly as much. I explained to them that when they are not getting as much money for their cotton, the denim mill will be making more money. So it brings the price of their cotton up to a higher level." Martin's economic theory may not have been sophisticated, but he "just stumbled my way through it. We got through. And we tried to answer their questions to the best of our knowledge, but always adding a disclaimer because we really didn't know how it might be."

To his complete surprise, the Hale Center Coop Gin Board voted then and there they wanted to be a part of the denim mill pool. Martin was more alarmed than pleased. He said to Cannon, "Well, Jay, really, you need to bring this to your membership." He was afraid the board did not completely understand what it was getting into. The 100 percent concept demanded overwhelming support from the membership of each gin. Even if the measure passed, a close vote could be fatal to the gin. It could mean the loss of considerable business from those producers who would not participate and would no longer be permitted to gin there. "And here they were," says Martin, "ready to vote for it without a membership meeting."

He told Cannon that if he were in Cannon's position, "There's no way I would take that responsibility on myself," and advised a different approach. "[Have] your board recommend unanimously that the Hale Center Coop Gin participate in the program. Call a membership meeting and explain it to them. Let one of us come out and explain it, and hopefully we'll do a better job than Roy and I did tonight."

That is what happened. And the membership of the Hale Center Coop Gin voted nearly 100 percent to participate, making theirs the first gin in the pool. Not all the meetings went as smoothly. The fact that farmers who chose not to participate in the pool could no longer gin where they had in the past was highly controversial. It ran counter to the independent nature of the farmers and it rubbed nerves raw. Dan Martin remembers an, ". . . old man one night who came up to me after a meeting and he said, 'Son, I'm going to tell you something. Ever since this coop gin has been here, I've ginned my cotton here. And ever since PCCA has been there, I've marketed my cotton through PCCA. But the day you tell me I've got to, you can go straight to hell!'"

At the time, there were about 120 gins on the High Plains and Rolling Plains of Texas. Martin says that he, Dan Davis, and C. L. Boggs would go over lists of gins that were about to hold membership meetings. "We would try to get them to let us meet with the boards first and then the members. We picked the gins we thought would sign up."

L. C. Unfred recalls there were some gins "that wouldn't even let [PCCA] come present it to their board of directors, let alone their entire membership."

There was no time to go out as a team, so Davis, Boggs, and Martin divided their effort, each selecting gins to approach. "Dan [Davis] would pick those he thought he could talk into the pool; then C. L. would pick those he thought he could bring in. Then I would get the rest," Martin recalls. He had been on the PCCA staff only since January, 1974. As the "new kid on the block," he got to meet with many of those gins that were thought to be hopeless. "But," he says, "I was doing fine at signing up gins on my own. The others couldn't figure it out. I told Dan Davis one day, 'Maybe it's because I speak their language. They understand me better than they understand you and C. L.'"

Dan Martin is a former high-school football player from O'Donnell, Texas. He grew up in cottonfields and gins, and he has weather-beaten skin and a perpetual tan. In contrast, Dan

Davis looks the picture of refinement, and C. L. Boggs, although rugged, looks the executive. Dan Martin is hardwood to their leather. And the differences aren't just superficial. Martin is one of the few managers on the staff of PCCA who was a "cotton man," not a marketer but a grower and ginner. As field director for PCCA today, he is perfectly cast. The farmers know him and his background, and they are comfortable with him.

On October 15, Dan Davis wrote to Ralph Griffitts of FmHA, telling him that, "Many producers in West Texas have already indicated their desire to participate in the textile mill and the seasonal pool marketing program." Perhaps that was not entirely accurate. Even as Davis wrote those words, the frenetic search continued.

The next day, Murrell Rogers wrote again to Lynn Futch, saying that, as of that day "some fifteen gin communities have voted to join the pool bringing the commitment to 150,000 bales. They have until November 1," he said, "to make their commitment in the West Texas area. . . ." With two-thirds of the clock having run down since the PCCA staff began to recruit gins for ACG, they were only three-fifths of the way to their target. Rogers continued, ". . . they expect some 30 gin communities-members to come into the pool by [November 1], bringing the total to approximately 250,000 bales. These gin communities will represent some three thousand cotton producers in the West Texas area. [Furthermore],

1. The farmers will put up fifteen percent [down from the 20 percent initially demanded] of the capital needs of the mill project in cash equities. Based on $30,000,000 capital needs, this would be $4,500,000.
2. The Bank will furnish eighty-five percent of the capital needs [$25,500,000] on a fifteen year term loan.
3. At least $10 per bale [$2,500,000] of the fifteen percent equity will be furnished by individual producers pro rata to their production.
4. The balance of the fifteen percent equity required [$2,000,000] may be supplied by purchase of equities in the association for cash by the other cotton regionals in the Lubbock area.

Hoping to prod FmHA into making the loan guarantee commitment, Rogers added, "We are pleased with the progress made by the cotton producers in the West Texas area in their development of this cotton marketing pool and the cotton mill project."

A few days later, Futch sent copies of Rogers's letters to Elliott in Washington. "The marketing pool," he explained, "is an essential component of the proposed financial arrangements, and, as explained by the Bank, adds substantial financial strength to the proposal as an alternative to Item 3, of the conditional commitment." He emphasized that debt service would take priority in any monies received through sales of cotton or cottonseed from the pool. Thus, the bank would be covered and, behind it, FmHA. He ended his letter to Elliott by saying, "we recommend your favorable consideration and acceptance."

CROSBYTON COMMITS

On October 28, ACG held its 1974 annual meeting. Since its founding, ACG had only one division, Crosbyton. The directors and members of CGD were the directors and members of ACG. In the textile mill pool, CGD now would become one of several divisions of ACG as originally planned, but not for the construction and operation of super-gins. Instead of being the nucleus of an eventual network of 15 or 20 super-gins, ACG now would become, at least in Dan Davis's mind, the nucleus of a network of textile mills. "The idea," he says, "was that we would build other textile mills, making other products, and that eventually we would have a one million bale pool, of which about 250,000 bales would be consumed in various textile manufacturing plants." Davis not only is creative, his mind is flexible enough to accommodate change. And the thoughts that poured from it never were small.

As in building support for the super-gin, strong leadership was essential. "It depended on the local leadership," Davis says. "If it was receptive to the concepts on which ACG was built, the community tended to come into the program. If they were negative, that persuasion would turn the tide. But we weren't looking for that much public acceptance. We wanted to get a small, tightly knit group of [leaders], like nine people. We didn't want it too big or too small. The idea was to get some sort of direction and then prove [our ideas] were sound."

The leaders of CGD–ACG had decided their cooperative should join the textile mill pool, but the members had not yet had their say. Two hundred and seventy-five members and guests convened at the October 28 meeting. The minutes of the ACG organizational meeting held May 4, 1973, were read and approved, almost 18 months after they were recorded. T. W. Stockton introduced Dan Davis, who gave a presentation on "the proposed cooperative textile mill."

After some discussion, Stockton called for a secret ballot on the Crosbyton Gin Division of ACG joining the textile mill pool. Ballots were distributed, and Stockton appointed a committee of W. R. Harris, C. D. Moore, and Gaylon Wheeless to count them. The result was: 86 votes for, 6 votes against joining the Cooperative Textile Mill. It was an overwhelming endorsement of the leadership's plans.

The rest of the meeting was routine. Two directors' terms were expiring — Charlie Wheeler from Crosbyton and G. B. Morris from Wake. Wheeler was nominated to succeed himself and elected unanimously, but Morris had asked not to be considered for reelection. The nominees for his position were Weldon Jones, Noble Hunsucker, Jr., O. D. Moore, Bob Ross, and W. H. Leatherwood. In a runoff between Hunsucker and Jones, Hunsucker won.

According to Martin, it was through the strong leadership of T. W. Stockton that CGD accepted responsibility for the denim mill. "[He] convinced the rest of the board over there to do it. And it's good that he could. We didn't have time to apply and get another coop charter and start over, because the FmHA funds were going away. We had to change horses right in the middle of the stream . . . change the name of the mill's builder to American Cotton Growers. We had to continue to meet with gins we hadn't met with. We even had to go back to some of the gins we already had met with and tell them, 'Hey, we told you wrong, we can't do this under PCCA. We're going to have to do it under American Cotton Growers.' Those were harder than the ones we hadn't met with at all." As Martin knows, "Cotton farmers pride themselves on being people of their word. When you go back to them with a change in your story, they start thinking, 'He may have been lying to me the first time.' You begin to lose your credibility. The first thing they would ask me was, 'What happened? The folks at [PCCA] decided they didn't want to risk their money?' It took a lot of reselling."

As the November 1 deadline approached, the PCCA staff had met with more than 100 of the gins on the High Plains and Rolling Plains. Davis, Boggs, and Martin each met with more than 30 gins during a five-week period. A majority of members at many of the gins wanted to participate in the program but were unable to because they could not have sustained the gin after those who would not participate took their business elsewhere. Davis, Boggs, and Martin told the gin boards, "Don't join unless 90 percent of your members are in favor. If you have even 60 percent, it won't be enough, because you will lose 40 percent of your members and it will break your gin.'' Not that every member voting ''no'' left his gin when it joined the denim pool. The experience of the communities that signed into the pool showed that about half of those voting against the proposal continued ginning there.

THE FINAL GIN

Twenty-six gins had been signed. One more was needed to reach the 250,000-bale goal before the FmHA commitment would expire, which, as Dan Martin recalls, ''was in about one or two days, and we already had talked with anyone who would listen.'' He describes the last meeting. ''I was meeting with Wilson Coop Gin at 10 o'clock one morning. Dan Davis was in San Francisco at the offices of Levi Strauss and Company. C. L. [Boggs] was in Houston at the Bank for Cooperatives. They were waiting for me to meet with this gin, the final one, and to call them and tell them that the gin either signed up or didn't.''

Martin recalls that about 80 or 90 members of the Wilson Coop Gin attended the meeting. It was held in the auditorium of a local school, and it took about two hours to go through the presentation and questions and answers. Then the members voted. Eighty-eight percent voted ''yes,'' they would join the denim pool. It was a solid, positive vote. But the matter still wasn't decided. After the membership vote, the seven-man board of the cooperative convened to take another vote, this one on whether the cooperative should join the pool when it might mean the loss of 12 percent of its members. They were empowered to override the membership vote if they felt losing that many members would imperil the financial health of the cooperative.

Martin paced outside the meeting room for more than 30 minutes while the board deliberated. The meeting ended. The door opened. Martin was invited in to hear the results. They had decided it was worth the chance — they would go with the wishes of the 88 percent and risk losing the others.

Wilson was in. The signatures were on paper. The job was done. Dan Martin got into his car and headed for Lubbock, 20 minutes away, where secretary Liz Haynes was waiting to place calls to Davis and Boggs with the news, good or bad. As he drove along the two-lane country highway between Wilson and Lubbock, the pressure that had built up over the previous five weeks burst.

He pulled his car off the road and broke into tears. In his words, he "cried like a baby." Two weeks later he had his first heart attack.

Dan Martin still gets emotional when he talks about the effort to sign the gins, particularly about the Wilson Coop meeting. But he laughs about it now. He recounts a trip he took with some of the Wilson Coop Gin directors in early 1988, more than 13 years after the denim mill vote. "Glynn Moore and I got to talking about it and laughing. Glynn was the one who made the motion to the board that they go ahead and join the program." Moore well knows the significance of the vote, but Martin reminded him anyway, "You were the people who put us 'over the top' and caused us to have a mill. Of course, you could have been the people who didn't put us over and caused us not to have a mill. Scary, isn't it?" Shaking his head, Martin repeats, "No, there would not be a denim mill. We would not have had time to set up more meetings, even if we thought there were more gins who might consider joining. And I don't think there were."

Building the Mill

SITE SEARCH

While part of the PCCA staff searched for gins to join the pool, another group, augmented with outside assistance, was searching for a site for the mill. A study team of five men was appointed, each responsible for researching, analyzing, and reporting on particular criteria. Emerson Tucker had utilities and fire protection; Bill Blackledge, taxes and transportation; Dan Looper, a consultant from Lockwood Greene, water and waste treatment; Dr. Louis Ponthieu, a consultant from Texas Tech, labor; and Jim Heath, from the Texas Industrial Commission, community incentives. From their areas of responsibility, the five developed a list of 120 detailed criteria by which the communities were judged. These criteria were assigned mathematical values, so the ratings could be derived as scientifically as possible.

Twelve communities that expressed an interest in providing the site were considered by the team. They were Brownfield, Crosbyton, Floydada, Hale Center, Levelland, Littlefield, Lubbock, Plainview, Seminole, Shallowater, Slaton, and Tahoka. Several communities fell off the list almost immediately. One, for example, did not have a railroad line. Several did not have adequate supplies of water or facilities for treating sewage.

When the data were all gathered and compared, Tahoka occupied first place, Littlefield second, Lubbock third and Levelland fourth. It appeared Tahoka would be the site of the mill. But that was only by the numbers. Several events took place that propelled Littlefield to the top of the list. Lubbock was taken off. In order to qualify for a loan guarantee under the Rural Development program, the mill had to be in a rural area, one under 50,000 in population.

George Mahon helped the city of Littlefield acquire an FmHA rural development grant to extend water and sewer lines to the mill site. That was a major attraction for ACG. Then the citizens of Littlefield raised more than $100,000 to purchase 105 acres of land for the project. That virtually assured their city would be selected.

The final factor, according to Tucker, was the most unscientific of all — what is often referred to as "gut feeling." The human factor in a process like this is nearly impossible to measure, but it is the one that most frequently turns decisions. Littlefield just seemed to want the mill more than anyone else. Tucker says that a couple of the communities gave the team the impression that "they didn't care that much." Through their intense efforts, the officials and citizens of Littlefield showed they cared a great deal.

With ACG not formally reorganized to take responsibility for the mill, PCCA was making the decisions. On January 13, 1975, the site study group reported its findings to the PCCA Seasonal Pool Committee. At the conclusion of the report, committee member Doyce Middlebrook moved the group vote by written ballot for the proposed site locations, ranking them in their order of preference. The motion carried. Ballots were distributed and the results tabulated. They showed the value of "gut feeling." Littlefield came out first. Plainview and Shallowater, which had not been even in the top four of the mathematical rankings, came out second and fourth, respectively. Tahoka was in third place. Levelland had fallen off the list. Immediately after the vote, a motion carried that the study group negotiate with the city of Littlefield to make arrangements for location of the mill there. If for any reason satisfactory arrangements could not be completed with Littlefield, then Plainview, the first runner-up, would be the alternative site.

Only the day before, the executive committee of the Seasonal Pool had recommended approval of a contract with Levi Strauss & Co. for the sale of denim fabric to be manufactured in the plant. The ACG Board, whose approval was required, affirmed it unanimously. Thus, in two days at the beginning of 1975, two major issues had been resolved — the location of the mill and the contract with LS&CO.

The performance of the cotton market had turned dismal. Consumers were not purchasing cotton products, so textile mills were not purchasing cotton. Dan Davis's assessment was that the "market is very bearish . . . in a very negative position." Despite that, he drove ahead on all plans for the mill. He knew that one of the most important elements, maybe the most important, would be selection of a plant manager. If ACG was going to build a denim mill, it would need a "denim man," a strong, experienced leader in the field who would oversee the project from construction, through production, to profitability. No one in the present scheme of things fit that description. ACG had administrators, marketers, engineers, lawyers, farmers, ginners, and visionaries, but no denim man. Not one. Even a pedigree in denim would not be enough. The cooperative decided the ideal candidate would meet three criteria. In addition to having 10 years of denim experience, he would be a native Texan and a graduate of Texas Tech. There could not be more than a few such men in the entire country; they were searching for a needle in a cotton bale.

THE "DENIM MAN"

Two thousand miles away, in Rock Hill, South Carolina, a denim man was in mourning. On February 4, Bob Hale's wife of 24 years passed away. Hale had been working for many years in the cotton and blended-fibers division of J. P. Stevens, one of the country's textile giants. A native Texan who had moved to South Carolina and married a local girl, Robert L. Hale now wondered how to go on without her. He recalls, "I was living in a small town in Carolina and kind of confined. I was living within a few miles of her mother and father and brothers and sisters and . . . all of her family . . . large and very, very closely tied, . . . and the idea of just going on didn't appeal to me. It was a real emotional problem." Within weeks of his wife's passing, Hale got a telephone call from Dan Davis. Davis knew his reputation in the textile industry. He also knew that he was a native of Texas and, coincidentally, an alumnus of Texas Tech. What he didn't know was that Hale was ready for a major change in his life.

Bob Hale pondered the invitation from Dan Davis to interview for the ACG plant manager's position, but he did not ponder long. "The idea of moving away . . . to a new location and kind of starting all over again had a real appeal to me. I knew that if I [did], I'd be meeting new people all the time, and it would be a challenge. I would be forced to kick myself in the butt to get going every morning and have a smile on my face."

What was happening in Lubbock was news to him, but not a complete surprise. Months before the call from Dan Davis, Hale had been contacted by Emerson Tucker, a fellow Texas Tech alumnus. The two had known each other as students. Tucker said he was doing some experimental work in open-end spinning and asked if Hale would manufacture some cloth from open-end yarns for him. Hale then was manager of J. P. Stevens denim plant at Rock Hill. He gladly obliged Tucker, and some open-end yarns spun at Texas Tech were sent to South Carolina and made into cloth samples. What Bob Hale didn't realize at the time was that he was making the original samples of denim from open-end spun yarn for PCCA.

Dan Davis's call began an interview process that was more grueling than Hale could have imagined. In February and March, he made "four or five trips" between South Carolina and Texas. "I would . . . catch a late, late plane Friday night in Carolina and fly out here and spend Saturday and Sunday and catch a plane back on Sunday night. After I'd done that about four times, I said 'Now, Dan, we've talked long enough. It's about time you're going to have to make up your mind because I just can't keep coming every weekend.'" He recalls Davis saying, "If you have a short interview with somebody, they have their guard up and they're just talking. If you stay with a man long enough, you're going to learn something about him. I'm determined to be with you enough time, enough hours, to get to know you before we make any decisions."

John Hunt of LS&CO recalls having cocktails at Dan Davis's home one evening during ACG's search for a plant manager and shortly before Hunt assumed new duties in Hong Kong. Davis sought his advice. "Who should we put in charge of this?" he asked. Hunt, well aware of "who is doing what in the industry," had contributed to the discussion of candidates. His answer was, "Bob Hale . . . good for PCCA, good for us." What made this endorsement significant was that Hale made denim for a company that did not even meet LS&CO quality standards. J. P. Stevens had never tried, because they sold their product to chain stores. Hunt knew little about Hale when his name first came up in discussion but had found out his reputation was solid.

Karin Hakanson of LS&CO knew of Bob Hale and had serious reservations, but not because of him. "I knew Bob Hale from [his work at] J. P. Stevens. He was in charge of their denim mill. And it was not a [modern] mill. It was a very old one." She was not certain that the manager of that mill could build and run one that would produce denim to LS&CO standards. Her doubts were heightened by her earlier concerns about the open-end spinning method. Today she gladly admits, "I was proven wrong on both counts."

Hale remembers the interview process seeming to reach the point of inanity. He was meeting with the ACG Board of Directors on a Sunday afternoon, and he was tired of being interviewed. Half in jest, one of the board members asked, "Well, do you like sandstorms?" Incredulous, he snapped back, "Hell no, I don't like a damned sandstorm." The board member replied, "Well, if you lied about that I guess you'd lie about other things. But seeing as how you told the truth on that, I figure you're truthful enough." Hale says, "They agreed to hire me right then and there." The decision, of course, had less to do with Hale's honest answer to that question than it did the final realization on the board's part that the process had gone far enough. They had found a denim man with a good reputation who also was a native of Texas and a graduate of Texas Tech. They should not take a chance on losing him.

Hale knows of only one other candidate for the plant manager's position, though he says there were others. Like Hale, the particular candidate he knew met all the criteria. Unlike Hale, however, he had a reunion with some of his "old Texas Tech buddies" the night before his first interview. The reunion may have gotten out of hand; the candidate never showed up the next day. Had the ACG Board lost Hale, it would have had a difficult time finding a suitable replacement.

By the time the interview process wound to a close, Hale was leaning toward taking the job if it were offered. He says, "I was in a situation where I only had the one thing [the job offer]. I mean I had no family situation, I had a daughter who stayed in Carolina in college and my son came with me out here. He was 14. It left me with one focus and that was the plant. I agreed that I would give two years of my life to build the plant. I knew that I would be just totally involved in it, and that it would be a wonderful way for me to start over again. . . . If these things had not happened, and I had to weigh moving a family, wife and children, it would have meant weighing job security [in South Carolina] against job security [in Texas] and it would have been a real problem. This wasn't a secure job." As an aside, he adds that getting his late wife to move might have posed an additional problem. "She was a very devoted South Carolinian. In fact, she told me that if I ever took her to Texas and buried her out here, she would walk back to Carolina. She just never loved the place, I promise you that!"

Sometimes he thinks about the remarkable order of events and the speed with which they took place. He says he doesn't "believe in fate and all that sort of thing, but still . . . I went to see my preacher because I was chairman of the administrative board of my church, and it was a large church, and if I was going to leave he needed to know. . . . I told him what the situation was and he, in the normal response of a minister, asked, 'Have you prayed about it?' And I said, 'Well, yes, I have. And as far as I'm concerned, the good Lord told me to go to Texas.' I'd just begun to feel a really peculiar situation, being drawn out here with everything falling into place for me to come out here. So I packed up."

"Gone to Texas" were words that Americans seeking a new life had said for more than 150 years. This time, Texas was bringing home a native son who would be instrumental in building a business epic.

THE RELUCTANT TEXTILE MANUFACTURERS

In February, 1975, it had been eight months since Dr. Bostic had said USDA would not proceed with PCCA's application until after the government complaint against it was resolved. And ACG still had not adjusted to the fact that it was going into the denim business.

The ACG Board meeting minutes of February 25 say that "Mr. Davis, Mr. Blackledge, and Mr. Boggs gave a presentation of the structure of the Cooperative Textile Mill as it *would* be *if* built in American Cotton Growers [authors emphasis]." After some discussion at that meeting, Clyde Crausbay pressed the issue on behalf of the 27 pool gins. He moved that, "the cooperative textile mill be built in American Cotton Growers; that the 250,000-bale pool from the 27 gin members of the mill be in ACG; and that a marketing agreement be drawn up between ACG and PCCA to market for ACG that cotton not used in the mill." The motion was seconded by J. K. Edinburgh. It carried, five votes to one.

ACG, officially at least, had embraced responsibility for the denim mill. If it seems the cooperative was reluctant to take that step, it probably was. Far more was involved than a simple change of name, from PCCA to ACG, on a sheaf of papers. A great deal of exposure to risk changed hands, too. When PCCA planned to build the mill, the Crosbyton Gin Division of American Cotton Growers planned to participate, but only as one of many

PCCA gins. Under the original scheme, PCCA would have shouldered the risk by pledging all its assets, including the denim mill and the compresses at Sweetwater and Altus. When ACG agreed to use its cooperative charter to apply for the federal support, it put its own gin "on the hook for the note," as Dan Martin says.

March, 1975, saw activity shift into even higher gear. In addition to hiring a manager for the mill, the ACG directors were being advised by Bill Blackledge on the steps they needed to take to complete the organization of the textile mill in ACG.

Bob Hale returned to Lubbock permanently on Easter Sunday, March 30, 1975. ACG announced his appointment the following Thursday, April 3. On April 4, the *Lubbock Avalanche-Journal* trumpeted his arrival with a headline story in its "Agricultural Journal" section. Although his arrival was roundly cheered, what Hale found when he got there was less than encouraging. It wasn't the physical setting that bothered him, though he recalls a visit to Littlefield during one of his interview sessions that caused him to wonder. "It was February, and it was a cold, bleak, windy day when we drove over to Littlefield and I frankly thought it was the most desolate ride I had ever seen in my life. You begin to shake your head and say, 'What am I doing here?'" He found unsettling some comments his PCCA hosts made when they toured the plant site that February day. "There was nothing out here but the land, and they looked at it and gestured and said, 'We'll probably have cotton receiving on this end, and there's enough of a drop, a four-foot drop, that we can have loading docks down at the far end." Hale listened to the words, but it was as though his eyes were seeing a different piece of terrain than the one being described. "I said, 'There's absolutely not a four-foot drop on this piece of land. It's the flattest piece of land I've ever seen!'" They were seeing what they wanted to see. More disconcerting was that the new job he had accepted seemed at times like a mirage. He felt very lucky, having been "given 105 acres of land and $40 million [*sic*] to build a plant." But after moving himself and his son west and focusing on the challenges ahead, he could see that "things weren't quite together."

Hale remembers one of the first board meetings he attended. In addressing the group, Dan Davis offhandedly said, "Well, after we get our loan finalized at the bank. . . ." After the meeting, Hale took Davis aside and said, "Dan, did I hear you correctly? You said *after* you get the loan finalized?" There was a painful silence. Hale, who had changed his entire life in response to Dan Davis's phone call, continued, "Dan, you ain't got no money, have you?" More silence. Then, "Well, no, we haven't, right now. But it will come through." And again, as if to reassure himself, "It will come through."

Weeks later, Hale was introduced at a meeting of the Littlefield city fathers by Bill Blackledge, who called him the "prospective plant manager." He responded immediately, "Bill, for God's sake, let's get two things straight. For one thing, I'm not 'prospective' because I've already been hired. I *am* the manager! For another thing, I'm *here* ! Now if you think that I'm still prospective, I better get out of here. *I* think I have a job."

ACG had not yet made the psychological adjustment to being in the textile business. If it seemed uncertain even weeks after Bob Hale arrived, the city of Littlefield had no doubts at all. To the host city, the denim plant was a fact of life. Littlefield was being touted in some quarters as a major player in Texas industrial development. A front page article in the *Lamb County Leader-News* entitled, "Denim Spinning Mill Makes Littlefield Industry Leader," referred to a Texas Industrial Commission report citing the city as the "industrial leader in Texas for announced expansion during March."

Reading the words of the chairman of the Littlefield Chamber of Commerce and Agriculture as he cited the TIC report, it's a little hard to tell that the mill existed only on paper. As far as announced expansion, the exuberant chairman said ". . . Littlefield not only had the largest single plant, but we also had the largest expansion on a city-to-city comparison. . . . I believe that's something to brag about when one compares us to places like Houston, Dallas, and San Antonio." It sounded like just a matter of time before Littlefield would be hosting a World's Fair. And all before the first spadeful of dirt was turned.

STARTING FROM SCRATCH

In May, 1975, ground was broken for the mill. C. L. Boggs was able to report to the ACG Board that 284,000 acres of cotton had been signed into the Textile Mill Division cotton pool. These acres were expected to yield 294,000 bales of cotton, well in excess of the bank's 250,000-bale requirement. Boggs's responsibilities with ACG had been building steadily. The CEA suit against PCCA caused the two associations to assume more of an "arm's length" relationship than they otherwise might have. Although Dan Davis was in fact the general manager of both, he was officially the general manager only of PCCA. And as time went on, he allowed Boggs to play an increasingly active role in ACG.

ACG adopted a district breakdown, placing its 27 ginning communities in seven geographical areas: District I — Amherst Farmers, Anton Producers, Littlefield Farmers, Shallowater Coop, South Smyer; District II — Friona Farmers, Hale Center Coop, Tulia Coop, United Farms; District III — Lea Coop, State Line, Windham Coop, Yoakum County; District IV — Tahoka Coop, Tahoka Farmers, Wells Farmers; District V — College Avenue, New Home, Wilson; District VI — Buford Coop, China Grove Coop, Glasscock County, Inadale Coop, Roscoe Farmers, Colorado City. District VII was the Crosbyton Gin Division, and it would remain known as that. It consisted of the three original ACG ginning communities of Crosbyton, Wake, and McAdoo. With two gins in Tahoka, ACG had 28 gins in 27 communities.Within weeks of breaking the gins into districts, the board began planning for an annual meeting of all patrons of ACG, new and old, on Thursday afternoon, June 12, 1975. On June 10, the Crosbyton Gin Division Board met for the last time as the board of directors of American Cotton Growers.

At the June 12, annual meeting of the reorganized ACG, T. W. Stockton introduced guests at the head table, including L. C. Unfred, the president of PCCA; members of the Textile Pool Executive Committee; and Dan Davis. Stockton presented his report to the membership and welcomed the new members who had joined to make the denim plant possible. One of the first orders of business was the establishment of an ACG Board of Directors. Nominations were received from district caucuses, and all the nominees were elected by the membership. They were: District I — Doyce Middlebrook of Shallowater, District II — Joe Leach of Plainview, District III — Dallas Brewer of Denver City, District IV — Harold Barrett of O'Donnell, District V — L. C. Unfred of New Home, District VI — R. H. (Billy) Whorton of Roscoe, District VII — T. W. Stockton of Crosbyton.

Stockton presented Roy Davis with a denim suit manufactured by Levi Strauss & Co. as a gift from the association for "his work in the development of textile manufacturing as a viable industry for the High Plains of Texas." It was good that Roy Davis was so honored. It would be one of the best and last tributes of his life.

Then it was time to show the members exactly what they were building. Dan Davis introduced Bob Hale, who showed slides of the textile mill project under construction. At the end of his presentation, Hale introduced his assistant manager, Fred Eddins, and the personnel manager of the textile mill, Mickey Brewer.

Eddins and Brewer were among the first on the plant staff. They were hired in the spring, the hiring process having been made easier for Bob Hale by the gentler weather. In reminiscing much later on his own hiring, Hale recalled that "February is a very poor interviewing month out here. And I assure you we never tried to hire any people from South Carolina to come out here to work in the month of February until [the site] looked better. To get a person out here in February to look at a job meant they weren't coming and that's all there was to it."

Finding other senior managers to share his West Texas experience was only one concern Hale had in the months after he signed on. More than any other person, he was responsible for bringing to fruition the years of hard work and meticulous planning by Roy Davis, Dan Davis, Emerson Tucker, Bill Blackledge, Joel Hembree, Jimmy Nail, and many others. He had to supervise the construction of the mill. He had to oversee the implementation of a complete denim production process, including equipment. "Now, I did not lay it out in detail for the construction," he says. "We were working with . . . Lockwood Greene. They were the textile engineers. We began looking at the layout we wanted. I had a lot of input of exactly what I wanted because I had more experience making denim. Emerson did not have the experience making denim, but he knows textile machinery, so he selected the machinery. Our first thing was to finalize a blueprint of a layout of the machinery and the flow of stock going through. Once we had that done, then the engineers took over and designed the building and worked out all the details for construction."

Hale feels ACG acted wisely when it hired a local company, Stringer Construction, to build the mill. Stringer, which was accustomed to building on West Texas terrain, worked well with Lockwood Greene, which was not. He remembers some of the problems Lockwood Greene had coping with this new and strange environment. "We had a hard time convincing them that we needed filtration on outside air coming in. They said, 'You filter the inside air that has the lint in it.' And we said, 'No, you filter the outside air that has the sand in it.'"

Hale also had to hire and train a full workforce, probably his biggest challenge. Even the newest and best equipment will not be effective without trained, dedicated operaters. A great deal had been done before he was hired, but much remained. And it was his to do.

Not inconsequentially, there was the matter of determining with certainty whether West Texas cotton could be turned into denim fabric of high quality using the open-end spinning process. Some denim manufacturers already were beginning to adopt the open-end technique. The question that loomed in front of ACG was whether the "trash" cotton of West Texas could be open-end spun into yarn that would make fabric meeting Levi Strauss & Co. requirements.

When Emerson Tucker had called Bob Hale at J. P. Stevens months earlier, Hale had some experience with open-end spinning. While at Stevens he already had conducted tests of the process and the yarn produced from it. "We had learned the quality of [open-end spun] yarn. We learned its characteristics." The tests revealed that yarn spun using the open-end method "just doesn't process like other yarns," Hale says. There were many technical aspects of the yarn, differences between it and ring-spun yarn, which had to be learned in order to process it through a plant and make denim from it. The implications for the production facility were enormous. Testing was a constant requirement. Unfortunately once Hale severed his ties to J. P. Stevens, the simple act of testing was a major obstacle for him and ACG. Even the act of visiting other textile facilities became a challenge.

Slamming Doors

The textile industry is a relatively small "family" in which the relatives all know one another but compete vigorously. They zealously guard their own facilities, processes, and markets. When the cotton farmers from Texas and their new denim man started to look for places to test their process, doors all over America slammed shut, particularly in the textile Southeast.

Hale says "We did a tremendous amount of traveling. If we were looking at buying a particular [type of] equipment, then we made trips through the textile Southeast to see installations. We wanted to see the machinery. You didn't just buy it out of a catalog." The reception Hale and Tucker received was none too warm. J. P. Stevens, for one, was off limits, despite the fact that Hale left "under the best of terms. [After I gave notice], they asked me if I would work something like four weeks, which,'' he explains, ''was unheard of at Stevens. Once you told them you were going, you were always gone. There was no severance pay or anything involved when you left Stevens. . . . It was a compliment that they asked me to stay. I had the very best working relationship with them and the best situation of leaving. But they had a policy that did not allow men of the competition to visit the plant. Period. They just didn't. So I didn't visit them again at all.''

Ironically, Hale may have been partially responsible for the policy being as rigid as it was. ''When I was at Stevens,'' he explains, ''my vice-president called me and asked me if I would show a person through the plant. He was a jeans manufacturer in El Paso and was planning on building a denim plant.'' Hale complied with the request. The visitor was Willie Farrah of the well-known Farrah Slacks concern. ''When the hierarchy in New York found out that I had shown our operation to Willie Farrah, everything hit the fan. An edict was passed that we will not show our facility to *any* of our competitors. It just drew the line.'' He smiles and says, ''So that's the reason I was never able to get back into Stevens.''

Stevens was not the only firm that turned a cold shoulder to Hale and his new employers. He tells a story that describes the attitude of the manufacturers and the experience of trying to get help from them. ''There was a very good friend of mine; at one time we were across-the-street neighbors in Anderson, South Carolina. A few years later he was responsible for a Swift Manufacturing operation. He was a hell of a nice fellow,'' Hale remembers. ''Swift was one of the largest denim manufacturers, and we were hunting all the information we could get.'' Hale's friend let it be known, through third parties, that ''I think Bob Hale is a great guy; he's a good friend of mine and I like him. But if he thinks I'm going to help him start up a denim operation, he's crazy as hell.''

How little help would ACG get from the industry? Hale says, ''We had to have an analysis of the waste water that was going to come from the plant, so we would know how to handle it in connection with the waste-water treatment facilities of the city of Littlefield.'' Part of the agreement between ACG and the city called for Littlefield to handle ACG's waste water after some pretreating of the water by the denim mill. One of the reasons the site had been selected in the first place was its proximity to

Littlefield's waste water treatment station. "We *had* to know," Hale says, "and [Littlefield] *had* to know what the waste water would be like, so we would know exactly how to treat it. We needed a bucket of waste water from an indigo-dye plant. Would you believe that none of our good competitors would even give us a bucket of waste water?"

ACG solved the problem by going outside the country. "We literally sent a man to Canada," Hale says, "and our friends at Dominion Textiles, where I had visited, gave us some waste water. We took it to Clemson University, and they analyzed it. "So if anyone asks me if our competition helped us, the answer is 'Hell, no!' They wouldn't even give us a bucket of waste water."

Where ACG was able to get a foot in a competitor's door, it generally was through the manufacturer of machinery they were considering. "For example," Hale explains, "the manufacturer of Picanol Looms would have an installation, and they would have built up a good reputation there with the [textile mill], and they would not only get us access, they would take us." But ACG's access would be limited to the specific area in the plant where the machinery being observed was in use.

In talking about the effect of competition on testing, Emerson Tucker recalls, "Our being able to do anything at all in the United States was curtailed. We couldn't dye fabrics, just couldn't do anything." And because ACG was able to do so little in the United States, Tucker thought, "Let's go south."

A discussion Tucker had with George Babcock, then manager of the Growers Seed Association, pointed the way. Babcock knew some textile people who knew of a mill in Mexico, the Compania Industrial Rio Bravo, S.A., of Chihuahua. The company was owned by two brothers, Luis and Tomas Alvarez, and they welcomed the cotton growers as strongly as the American mills had rejected them. Tucker has maintained contact with the Alvarez brothers over the years, long after they sold their business. "They were lovely people," he says, "very accommodating. They are sharp people, too." One of them had been a candidate for the presidency of Mexico.

Finding a mill willing to produce fabric samples for ACG meant that all the pieces necessary for conducting comprehensive tests were in place. Bob Hale describes how the process worked. "Texas Tech spun the yarn from blends of cotton that we classed on high-volume instruments at PCCA. They were blends that, according to Emerson and Joel Hembree, would give us enough fiber strength that we would come out [with the desired fabric quality]. Then we shipped the yarns down to Chihuahua. . . . I made several trips down there as we processed the yarn through dyeing, beaming [process by which each strand of yarn dyed in a rope form is separated, straightened, oriented, and wound onto a beam], and weaving, and it was there we were able to get out samples so that we could get our strength checked and get some idea of fabric assurance.'' From Chihuahua, the finished fabrics were sent back to ACG, then on to Levi Strauss & Co., where they were strength-tested and evaluated.

Tucker recalls that, although LS&CO ran tests and provided feedback, ''we [tried] to zero in on a fabric that would be [good] for them and also would build a little confidence that we could, in fact, do what we had hoped to do.'' The determined efforts of a university, a cotton marketing cooperative, a Mexican textile plant, an apparel maker, and a bunch of cotton farmers were not to be stopped, no matter how many doors slammed shut.

Bob Hale's cooperation with ACG when he was still with J. P. Stevens was an anomaly in his industry. But he found the same attitude among others when he arrived at ACG, one perfectly appropriate to the cooperative spirit and openness of West Texas. ''We began with a strictly open-door policy,'' he says. ''We opened our doors to our competition. They could come in and see what we were doing.'' The ACG Board felt strongly that this was the best policy, and not just because they are friendly people; there was sound business logic involved. ACG reasoned that any opportunity to show visitors what the plant was doing with West Texas cotton would be good for sales. Increasing sales of West Texas cotton would benefit everyone in the region – ACG, other cooperative producers, and independent producers.

The open-door policy has not hurt. Hale says it has helped. ''The vice-president of manufacturing for Cone Mills, the supplier of LS&CO's 501 fabric, came out after we were in operation. He told me later that he came out because his board of directors had asked him, 'What about that bunch of farmers out there in Texas that are building a denim mill?''' The executive said he went back to his board and ''. . . told them that you were for real, that you were going to be there, so we might as well be on a friendly basis with you.'' Hale adds, ''From that point on, we were able to visit Cone and Cone visited us.''

Burlington Industries, another major denim manufacturer, never let ACG into their plant. ''From day one until now, we've never been able to get into a Burlington plant to see their operation," Hale says, but in fairness adds that their policy is consistent. Burlington won't let *any* competitors in their door at all. Nonetheless, Burlington has visited Littlefield. By participating in a seminar at Texas Tech that featured a tour of the plant, Burlington representatives saw what ACG has been denied, a full view of the competition's operations. "I permitted the seminar group to come over here," says Hale, "not knowing everyone that was in the group." And although it sounds as though Hale would have denied them access to the plant had he known who they were, it is highly doubtful. Denial is not Bob Hale's style.

"We take a completely different philosophy from the rest. We open the doors and, believe me, we get more information than we give. There's no question about it. I can pick up [the telephone] and call Greenwood Mills and say 'I've got a problem on the finishing range,' or something about shrinkage, and they'll tell me everything that they have about it because I've told them the same about other things. . . . I have given them samples of my fabric. They said, 'Would you give us an analysis of your fabric?' I said, 'Hell, no. I'll give you a sample of my fabric. You can analyze it any way you want to.'" He adds, "But there's nothing secret about that. They can go to Levi and say, 'Hey, we'd like a yard of fabric out of [ACG],' and they'll give it to them. I can go to Levi and say, 'I want to see some of our competition's denim' and they'll give me five yards of it."

Hale scoffs at the industry attitude. "It's really stupid, when the hierarchy think that they are protecting themselves so well. Those of us manufacturers who work together like to say, 'Let the salesmen fight it out as to who makes money.' But as far as making a good operation, we all benefit by exchanging information, and we do. So, with *all* the denim manufacturers, we've had an open door and it's worked out well."

With the new board representing all the pool gins having been elected at the June, 1975, annual meeting, it was time in July to elect the association's officers. By secret ballot, Middlebrook, Unfred, and Stockton were nominated for president. Unfred won. He then distributed ballots for nomination of vice-president. Stockton, Whorton, Leach, and Middlebrook were nominated. Stockton won. On the management side, Bill Blackledge was nominated for secretary and won by acclamation, and the same was true for C. L. Boggs as treasurer. The team was in place. Although Dan Davis was not officially affiliated with ACG, he continued acting as general manager. He ran the staff for ACG, just as he did for PCCA.

Climbing Costs

As the cotton market was going down, and even before the walls of the mill went up, Dan Davis showed his aptitude for daring. He told the board, almost offhandedly, that the plan for the mill had been expanded to permit an increase in production. The revised plan called for enough space to install 408 looms instead of the 340 looms in the original plan. Davis said the future installation of this equipment would bring plant production up to the capacity of the indigo-dye range, which would have a somewhat greater volume capacity than originally planned.

It may be true that the capacity of the indigo-dye range was driving the decision to increase loom capacity and production. It also is possible that a larger mill than the one planned was desired all along. Bill Blackledge indicates the latter is true. "When Emerson and Dan went to Europe to this textile machinery show," he says, "they saw the prototypes of these new 'shuttleless looms.' So they came back and redesigned that whole weaving-room operation and enlarged it so that it would accommodate those kinds of looms. [The manufacturers] wouldn't sell us the looms then because they weren't quite perfected. So we bought the old-style [shuttle] looms, but we put them in positions where we could replace them with shuttleless looms."

The manufacturer might not sell ACG the new looms at that time, but even if it would have, it is highly unlikely the ACG Board would have approved or the bank would have financed the purchase. Dan Davis seems very deliberately to have decided what course the denim mill would need to take in the future and committed the organization to that course without its knowledge or that of its bankers. "Nobody knew that we had done that," Blackledge says with a mischievous smile. "It really raised the price of that mill considerably." He is right. The cost overruns resulting from it nearly killed the association. But the added capacity and higher quality that Dan Davis envisioned, and that actually resulted later, helped bring the plant to the forefront of American denim production.

With an air of sincerity, Davis admits, "The thing we did a very poor job of controlling was the cost of building the plant itself. We built the building on a 'cost plus fixed fee' arrangement, and our main interest was to try to get it done as quickly as we could, to get into production while the denim market was hot and you could sell everything you could make, whether it was good quality or poor quality. "We spent," he says, "a lot more money building the building than the budget on the project. We increased the size of it about 15 percent during that phase. We had a cost overrun of about $5.5 million, which caused great consternation, and rightly so." The cost overruns were beginning to be seen as early as July, 1975. Davis told the board the HBC requested that ACG file a supplemental application for an additional guaranty commitment, "to cover the present anticipated costs and possible cost overruns on the project. Present projections," he said, "indicate that fixed assets for the mill will cost approximately $32,000,000. Approximately $2,000,000 will be needed in working capital and $1,721,000 for pre-start up costs."

T. W. Stockton moved the board approve the filing of the application, with a 15 percent reserve for contingencies on equipment changes and inflation and a 10 percent reserve for possible foreign currency fluctuations. The motion was seconded by R. H. Whorton and carried unanimously. The board voted to approve the filing of an application for a loan of $33.5 million with the HBC. It then considered issuing building bonds for the textile mill project. (It is interesting to note that the minutes of the board meeting actually say the "proposed" textile mill project, as though the board was not quite certain in July, 1975, whether it was going into the venture.) It authorized a total of $4,982,510 in building bonds for the project with $3,000,000 to be issued to PCCA for the regional cooperatives' investment and the remainder to be issued to participating textile pool gins.

In Littlefield, applications were being accepted for employment at the plant. ACG set up a temporary office in a storefront, the site of a former furniture store at 229 Phelps Avenue. The plant's first pieces of machinery, six Picanol 69-inch looms, arrived from Belgium and were installed at the storefront for training weavers and loom technicians. Twelve supervisor trainees would be schooled on erection, operation, and maintenance of the looms by Bernard Aerdker, a Belgian in the employ of Picanol. Aerdker had begun working with looms at the age of 14 and was considered one of the best at his craft.

Within a few months, the six looms were joined by other equipment. Bob Hale recalls the storefront operation. "We put up a drawing-in frame where we had people who were drawing the original patterns to go on the loom. [Drawing-in is a mechanical process by which each strand of yarn is run through all components of the loom in preparation for the loom beam.] We put in beaming frames that showed how we would beam the yarn. This was yarn that was dyed in Mexico that we were bringing up here. We actually began our initial training at the storefront for our beamer tenders, drawing-in people, weavers, and loom fixers. The door was always open. People walking down the street were welcome to come in and see the operation. We set up our employment office there. It [was] our policy from the very beginning to be as open as possible and let everybody know about our operation, and believe me, I found out that's the way to get along with farmers, to be completely open and honest.'' Not all the training going on Phelps Avenue was technical — supervisory training was provided by personnel from South Plains College. Other classes were taught by Franklin Sears, a labor attorney, and a management consulting firm called Berkman & Associates, Inc.

A few miles outside Littlefield, on State Road 54, a large sign went up on a tract of featureless dirt. In large letters, the sign proclaimed ''West Texas Cotton,'' and underneath those words, in smaller letters, was what seemed at the time more a hope than a promise: ''On this site, American Cotton Growers will produce 20,000,000 yards of blue denim per year with West Texas cotton.''

In August, construction at the site was slowed by heavy rains. Walking through the thick layer of mud left by seven inches of rain, Hale muttered, ''I thought this was semiarid country.''

August brought tragedy to the Lubbock cooperative family. That was the month in which the Plains Cooperative Oil Mill nearly failed financially from its ill-fated entry into sunflower-seed contracts. On August 20, there was a joint meeting of the executive committees of the oil mill, Farmers Cooperative Compress, Plainview Cooperative Compress, and PCCA, as well as the ACG Board. Joe Rankin, president of the oil mill, announced his executive committee had met earlier that day to discuss financial problems of the mill and decided to terminate manager John Herzer.

DEATH IN THE FAMILY

Four days later, the man more than any other responsible for the success of the oil mill and much of West Texas agriculture died. ''Mr. Roy'' passed away much too early, before his textile mill was completed, before the first denim was produced. The stately and esteemed gentleman, friend to the powerful and the powerless, champion of cooperative agriculture, was gone. His death came quickly. He was admitted to the hospital on Friday, August 22, after suffering chest pains. At two o'clock on Sunday morning, life left him. To the many who knew and admired him, it came as a complete shock.

His passing was reported on the news pages and lamented in the editorial columns of the local press. The accolades poured in, and rare was the one that did not in some way describe him as a visionary. An editorial in the *Avalanche-Journal* said he was "looking enthusiastically to the future up to the day he died. He had accomplished so much, but he was neither satisfied nor smug; it was what he could do tomorrow that counted."

Avalanche-Journal columnist Kenneth May expressed wonder and admiration over Davis's attitude. "To be looking around," he wrote, "with zest and enthusiasm, for new frontiers to conquer right up to the very day you draw your last breath: That is victorious living."

And Duane Howell, another columnist for the same newspaper, described him in words that are elegant for their simplicity. "Roy B. Davis," he wrote, "was good, he was modest, he was loving and humane. But his 74-year life left no doubt that he was also great. When he walked among giants, they often seemed a little smaller." So turned a page on Roy Davis, but in a book that remains open.

BUILDING THE TEAM

The "Dirty Dozen"

Several weeks later, in Littlefield, Danny Davis (no relation to Roy or Dan Davis) started work at American Cotton Growers. On the day he reported, which he clearly remembers as September 15, the ACG textile mill had five employees. They were Bob Hale, Fred Eddins, Mickey Brewer, Bobby Jones, who was a maintenance superintendent, and Cheryl Yarborough, who was secretary to them all. Davis, an engineer who had just earned a master's degree in business administration at Texas Tech, was one of the first two employees trained in the storefront. "Every other day after that, for about two weeks," he recalls, "they started two more people, so they kind of had us spread out. . . . They did that until we had the first 12 supervisor trainees who came to be referred to as the 'dirty dozen,'" after characters in a popular 1967 war movie.

Bob Hale remembers the "dirty dozen" all too well. ACG had decided to develop its own supervisors. It made sense to use local talent to run what essentially is that most local and personal of business ventures, a cooperative enterprise. "We started out," he recalls, "with a nucleus of 12 young men, all of whom had from two years of college to a master's degree. They were in the 28 to 35 age bracket. Just *exactly* the people we wanted to train." Hale pauses and a look of irony mixed with disbelief comes to his face. "We were so *stupid*," he says. "*I* was so stupid not to say, 'Hey, what are these guys doing on loose pulley?' as we say. In other words, 'Why aren't they settled in jobs?' Here were guys with good personalities, good education and just the right age. Why weren't they located?" He pauses, then continues, "To make a long story short, of those 12 people, I only have one left."

What happened to the others? "[Most of them] had sort of been misfits wherever they had been. Their qualifications all looked great, their energy was great. But oddly enough, most of them, after we had put them through a lengthy training program and got them on the job, would say, 'I don't want to supervise. I just don't want to tell people what to do.' They had been in positions [before coming to ACG] where they didn't have to assume a lot of responsibility and, when responsibility was pushed on this group, they melted. They disappeared. . . . One of them came back as an hourly employee; the other people are still around town here doing first one thing then another," Hale says. Sarcasm in his voice, he adds, "That was a *good* mistake we made." The one of the "dirty dozen" who stayed is Danny Davis, today the plant's personnel director. Hale calls him "a really outstanding man."

Danny Davis recalls the lengthy training program. In addition to technical skills, he and his fellow supervisor-trainees received instruction in management responsibilities. "The same day [I started]," he says, "another gentleman named Sid Mack started to work as . . . a training coordinator." Mack worked for South Plains College, which was providing training for ACG managers under an economic development grant. "Mr. Mack generally provided our training that was not related to the looms themselves. [We learned] how to deal with employees, supervise and train them, and those kinds of things."

When the local pool of supervisory talent was found to be wanting, ACG turned toward the area where it found Bob Hale, the textile Southeast. That was not a resounding success, either. "Frankly," Hale says, "the caliber of people that we could get to come to Littlefield, Texas, was rather poor. Most of them were running away from something . . . running from debts, running from a wife or family, running from the bottle. . . ." He strains to find the word to describe these early applicants, some of whom became employees. "Raunchy," he says. "We got a rather raunchy crew out here and that's the truth. They were hell-raisers and we didn't get the best reputation in the world. You know, a few fights in all the local joints and all that." Then he adds, "But intermingled with all those we got some pretty good fellows."

Bob Hale recounts some of the problems in training the workforce. "The state of Texas offered to help us train our employees by furnishing training manuals, paying for a director of training, and so forth. This was all good. We appreciated any money and help we could get . . . . The offer came from Austin, but the [state government] people there realized they didn't know anything about training textile people, so they hired the training expertise of South Plains College over here at Levelland. The training coordinator [Sid Mack] was on the payroll of the college. The college knew training, but they realized that they didn't know anything about textiles, per se. So they hired Arthur D. Little, a consulting firm, actually to put the technical information together. . . . Little knew textiles, but they didn't know denim, so they hired an ex-industrial engineer from Cone Mills . . . and so forth. . . . When we got our first training manual for looms, it said, 'To start the loom you grasp the shipper lever and pull it forward.' We said, 'Wait a minute! Our looms have electrical switches.' The manuals were that far off.

"Eventually," he adds, "we hired a man named Jinks Patterson who had been at Georgia Tech doing training work for them with the textile industry. We hired him as our training director, and he began to write our manuals and put together something we could get our hands on. Then we began training our own instructors to train new people."

Formulating other company procedures would prove no easier. "We had to start absolutely from scratch," says Hale, "on all company policies, on all pay scales, job loads, employee policies, vacations, insurance, everything, because we had no parent company to turn to for guidance." The task was less imposing than it might have been, thanks to personal contacts of the mill staff in the denim industry. "We got the full company policy from five major denim producers," Hale says. "Then we all sat around a table. We'd look through the manuals, and we'd bring up one subject.

Someone would say, 'This company does it this way.' Someone else would say, 'Well, this company does it that way.' We pulled what we thought was the best out of all these manuals, and we think we came up with a very excellent manual. We developed some innovative policies, and other firms have adopted them. We're proud of that."

But Can They Make Denim?

On October 23, John Hunt and Gil Melbardis of Levi Strauss & Co. visited Lubbock to talk with ACG officials and inspect progress of the denim plant construction. Hunt was on his way to Hong Kong to become general merchandising manager of LS&CO in the Far East, and Melbardis was taking his place as merchandising manager in the Basic Jeans Division. Melbardis recalls he was very interested in what he had heard about Littlefield, but like nearly everyone else at LS&CO, his optimism was tempered with some doubts. "My questions, basically, were ones like, 'Can I count on this production to cover our [apparel] plant? Can I actually schedule it that firmly, knowing there might be start-up problems? Can they meet our physical requirements using open-end spun yarn?,' and so forth."

Melbardis says, too, that LS&CO was somewhat concerned about the effect that its relationship with the West Texas coop would have on those with its established suppliers. "We were taking a lot of jabs from the older suppliers at a certain point. They were saying, 'Hey, this is a cooperative structure, and don't they have a different tax base? Don't they have an unfair advantage?' It was a bit of a concern to us, but the fact that somebody was willing to come in and actually make an investment in [a denim plant] when we were generally short of our denim requirements was all positive for us." The relationship LS&CO had with its older suppliers never was damaged. "They groused about it," he says, "but we all realize we need one another."

When Hunt and Melbardis visited, construction activity at Littlefield was showing signs of progress, with the administration building about three-fourths completed. About 95 percent of the foundation work was done, as well. One hundred and two of the total 362 Picanol looms had arrived and were being stored.

LS&CO officials were frequent visitors to the site, serving, in the words of Karin Hakanson, as "technical advisors." She, Hunt, Melbardis, McDermott, Pels, and others never tried to dictate what equipment a denim maker should buy, but they made themselves readily available to ACG and their other suppliers to give advice on process improvements that would enhance the quality of the product.

At the beginning, Tom Kasten of LS&CO had serious reservations about PCCA's lack of denim expertise. His visits to the site quickly dispelled them, and he became yet another strong advocate for the venture. Kasten had called some friends at J. P. Stevens as soon as he found out Bob Hale had been hired. He recalls they gave Hale "very high marks." On his second visit to West Texas, early in 1975, he walked around the foundation with Hale, who outlined for him, crisply and distinctly, where all the equipment would go and how it would be used. "You could hear in our conversation," says Kasten, "that he was well aware of textiles and denim. . . . There still were some 'doubting Thomases' [at LS&CO] at that point, who said, 'They're never going to be able to make a fabric that is consistently going to pass our standards.'" But their number was getting smaller.

Structural steel had been in short supply and was expected in early December. It arrived on time, and the first steel was erected quickly, though it came down even faster. On a December Saturday, a 75-mile-per-hour wind blew down or twisted whatever had been put up. The estimate of damage was between $10,000 to $25,000, and the estimate of lost construction time was two days. At this point, time was as important as money.

By the end of 1975, over 600 applications for jobs at the mill had been received. The hiring schedule was arranged on a week-by-week basis, contingent on completion of the mill. The training program was nearing the end of its first phase, with the second phase expected to begin in January. There also began to be a financial tightening. C. L. Boggs reported to the board that the availability of funds was getting "very short."

With the arrival of 1976 came the first expressions of organized discontent with the 100 percent concept. At its January meeting, the board considered a letter from the Lamb County Cooperative Gin, a non-pool gin. It was losing business because of the requirement that pool participants gin their cotton only at pool gins and requested the policy be reconsidered. There was discussion of the request, but it was brief. The board quickly and unanimously made clear its stand. On a motion by Dallas Brewer, it reaffirmed the conditions of its policy — all growers ginning at pool gins must be in the pool, and all pool participants must gin contracted cotton at pool gins. This policy would be reviewed almost constantly over the next two years.

At Littlefield, Bob Hale estimated denim production would begin around the first of May. Normally, the mill would have manufactured its own yarn using members' cotton. That yarn then would be turned into fabric. But delays in equipment deliveries would keep the mill from spinning its own yarn until sometime in June. In order to begin producing denim before the spinning operation was in place, Hale said, the mill would purchase yarn from outside sources. It was more important to begin making denim than to wait for all the pieces to be in place, even if it meant making denim from someone else's cotton.

Gil Melbardis of LS&CO would be the principal contact with ACG during 1976 and 1977. He paid several visits to Littlefield while the plant was under construction. On February 5, 1976, he addressed the Littlefield Chamber of Commerce. "On that particular trip, as I recall, they were just starting to put the roof up. There were ladders on the side, and I was asked if I wanted to look at the roof. I climbed the ladder and looked at the roof." He pauses. All he could see were a few steel beams, some portions of a roof beginning to take shape, and a vast expanse of Texas soil. He draws on a cigarette, grins a wide grin, and says, "Veeerrrry interesting."

Melbardis's thoughts were less on construction than on issues involving the quantity and quality of the material that ACG might be able to deliver. "Obviously," he says, "as you start scheduling your plants . . . you are, in effect, making plans for a year. You're saying, 'We're going to run so many sewing plants with so many operators,' and it's critical that supply be firm, or else you're shutting the plants down . . . and sending people home."

By early March, the Textile Division of ACG had 65 hourly and 17 salaried employees on the payroll. Danny Davis, who had been in training at the storefront since September, had moved to the mill around the first of the year. By the time he arrived there, "the concrete flooring had been poured and there was iron work started on the south end of the plant where they were erecting some of the steel tube and beginning to put in walls."

He remembers, "Some of us started coming out [to the plant] to be involved in things related to construction. . . . Some of the other people stayed at the furniture store, and we kept people over there in training in various ways for six or eight months into 1976 before we actually pulled everybody out and brought them over to the plant. . . . Most of the people we were hiring then were loom technicians and weavers, generally local people from about a five- or six- county area. People were coming in and putting in applications. To start with, we did get quite a few people from Lubbock, but especially from all the small towns around [Littlefield]'' — Anton, Spade, Amherst, Whitharral, and others. ''Most of them had only worked on farms or in stores, so they had to be trained from scratch.''

As the steel beams at the plant went up, the acreage of cotton signed into the pool went down. At the end of the April, 1976, sign-out period, Dan Davis reported that contracts on about 15,000 acres of cotton in the textile pool had been terminated. This included cotton acreage on which tenants had changed, acreage that might (or might not) be signed back into the pool for the next crop. At the time, it was hard to tell just what was on the minds of the farmers signing out of the pool. Everything was just too new and uncertain. In retrospect, it had much to do with apprehension about ''the way things were going,'' including the cost overruns at the mill.

As ACG's staff legal counsel, Bill Blackledge was responsible for enforcing the textile pool marketing agreements with members who failed to deliver their cotton. The board, in cooperation with the local gin associations, was determined to pursue legal action against defaulting growers. A deal is a deal. A contract is a contract. There would be no exceptions. But the problem did not always involve balky farmers who decided not to deliver the cotton they had contracted to the pool. Sometimes it simply proved impossible for a producer or a gin to stay in the program. Lea County Coop Gin was such a case. The gin had a problem because corn was replacing cotton as the dominant crop in the county. In April, the gin management communicated to the ACG Board that its continued participation in the pool would be difficult.

The developments were ominous. ACG had toiled to sign gins into the pool. Then there were expressions of dissatisfaction with the 100 percent concept. Then there was a small but troubling sign-out of pool acreage due to concerns over the mill. And now, the cooperative was in danger of losing one of its gins.

Jimmy Nail does not believe today the jeopardy in which the Lea Gin found itself was due completely to a change from cotton to corn. He contends the pool ''had a lot to do with it,'' and recounts a number of gins that found themselves in desperate financial straits that he attributes to the 100 percent requirement. ''The gin at Buford got in bad shape financially. Anton got in pretty bad shape." He mentions China Grove and Tahoka Coop, too. "They needed volume and [because of the loss of members due to the 100 percent concept] they just weren't getting it." He accompanied Bill Blackledge to College Avenue Coop Gin to confront one farmer who insisted he would gin there whether he was in the pool or not. He told them, "The hell I'm not going to gin here," and threatened to get a lawyer and sue the cooperative. Some of the gins never recovered, says Nail. "Part of it could have been from their management, part of it from economic conditions of the time, but the beginning of it was that situation right there. That's my opinion."

Despite these events, the board discussed the possibility of establishing a deadline for signing into the pool and formulating policies to control the growth of future patronage. It is not clear what led the directors to believe, at this point, that barriers would have to be erected to keep the pool from growing too large or too quickly. That would not be the problem at all.

Most attention remained fixed on Littlefield. A good description of the denim mill, replete with statistics, appeared in the June, 1974, issue of the magazine *Progressive Farmer*.

> West Texas farmers figure there's more profit in blue jeans than in cotton. About $30 to $50 a bale more. Not that they plan to go into the clothing business, but, beginning in a few weeks, growers will be manufacturing the basic denim material that has become the hottest selling fabric in the world.
>
> This is no namby-pamby deal. A group of some 3,000 producers is building a textile mill . . . that will take cotton they produce and convert it into denim. When operating at capacity, the plant will digest 183 bales of homegrown cotton per day and spit out 2,381 yards of fabric per hour.
>
> Everything about the project is huge. The plant – located about an hour's drive northwest of Lubbock – has 8.4 acres under one roof and houses the most modern open-end spinning setup in the world.
>
> Its price tag is big too. Projected at over $30 million, the cost is being underwritten by growers through the 26 cooperative gin associations that handle their cotton and three Lubbock-based regional cooperatives – Plains Cooperative Oil Mill, Farmers Cooperative Compress, and Plains Cotton Cooperative Association. Long-term financing is being furnished by the Houston Bank for Cooperatives.
>
> The farmers, operating under the banner of American Cotton Growers (ACG), produce about 315,000 acres of cotton in the High Plains and Rolling Plains areas of Texas. About 63,000 bales will be used in the textile mill. The balance will be marketed through a marketing pool. Weaving cotton into cloth is expected to increase its value to members by about $30 to

> $50 a bale. All members of the marketing cooperative will share equally in the profits generated by the denim plant.
>
> Texas-reared Bob Hale, manager of the textile division of ACG, explains that cotton bales will be brought into the mill where they will be opened, blended to make up a uniform raw material, cleaned, carded, drawn, spun, dyed, woven, and preshrunk. The finished product will be 14.5-ounce, indigo-dyed, white-filled, 100% cotton denim.
>
> At full capacity the plant will turn out 20 million yards of 60-inch-wide material a year. That's enough to make up about 19 million pairs of blue jeans — one pair for every man, woman, and child in Texas, Oklahoma, New Mexico, and Arizona.
>
> The open-end spinning process makes it possible to use short-staple, lower micronaire qualities which will contribute to wider markets for southwestern cotton.

The article reminded readers that, "the most remarkable fact is that all of this is owned and controlled by cotton growers. . . . They — through a seven-member board of directors composed of cotton producers — will set the operating policies of the mill. A 26-man committee [the Textile Pool Committee], representing the member gins, sets sales policy for the raw cotton pool."

In noting the success of the seasonal pool, it quoted PCCA spokesman Vern Highley. "As of April 1, [1976]," he reported, "cotton marketed through the pool was averaging nearly 2 cents a pound more than non-pooled cotton."

"Add to that the profits from the denim plant," said the article, "and you come up with a bonus that could approach $50 for every bale of cotton a member produces. That will buy a lot of blue jeans — with plenty of strong pockets to carry the extra 'jingle' in."

Before this optimistic view of the future would be realized, some obstacles would have to be overcome.

The Longest Yard

On May 6, the mill produced its first yard of denim. Bob Hale well remembers the event. "It was in May, and I was determined to have a yard of cloth for [the monthly board] meeting." He talks about the effort that went into that first yard. "We worked all night long. I went home around six or seven o'clock in the morning and got something to eat and shaved and cleaned up and came back. By that time, we had the first yard of cloth woven off the first loom. I took it to the meeting and showed it to the board. I said, 'This is my report. Now you know we can weave cloth . . . that it'll all work . . . that it'll all come together.' They were glad to see a piece of cloth and I was proud to be able to give it to them. No question about it. They realized . . . and I did . . . that the first yard of cloth cost us $17 million. That's what we had spent to get to that point."

When asked where the first yard is today, Hale says "I have it." After a pause, he grins and adds, "L. C. Unfred, our chairman, has it. I believe Dan Davis has it, and so does Emerson Tucker. That first yard of cloth is in about a half dozen hands, because we had a few more 'first yards' that we distributed soon after the original was produced, so everybody could have 'the first yard.'"

Bennie Williams, now a quality control lab technician at the plant, recalls the early days of denim production. "It was like working outdoors. Machinery was operating on concrete slabs, but all the walls weren't up. You had the wind and the dirt whistling through. And all the time, there were workmen overhead. We had to wear hard hats in case they dropped tools. And they dropped a lot of tools."

The hazards weren't just overhead. There were numerous tunnels under the plant for the installation of cleaning equipment, plumbing, and air ducts. Danny Davis remembers, "When we were making one of our first dye runs, the dye range was running at night. Several of us had been up all day and then all night while the dye range was making that run. It was about three o'clock in the morning, and we were all getting sleepy. We decided to come up to the front office to get a cup of coffee. . . . We left the office and were going through the weave room, back to the dye range. It was dark and we were kind of feeling our way along. There was an older gentleman by the name of Clyde Turner who was our dye range superintendent at the time. He was walking along with me and we were talking about how things had been going that night.

"Suddenly, Clyde wasn't there. Because it was almost totally dark, I had no idea where he had gone. I started calling, 'Clyde? Clyde, what happened?' I heard a few noises coming from below me, then I heard his voice say, 'I think I fell in a hole.' Sure enough, he had stepped into a hole about six feet deep and had totally disappeared. We all were a little more careful after that."

Hale talks about the push to get out the first yard of denim. "We had to break a few [construction] rules." In the winter of 1975–1976, "the weather was especially cold and one of the rules, I believe, is that the temperature has to be 40 degrees and rising before you can mix mortar so it won't freeze. We had some brick masons here who would throw water on the mortar board and if it didn't freeze, they would go ahead and mix it. They had to. It would be far up in the day before the temperature would be above 40 degrees, and we had to put up walls."

In West Texas, the wind is a constant companion. "It was blowing all the time," Hale says. "We realized when we erected some walls we had created wind tunnels, which increased the velocity of the wind. We would put in sand to be ready to pour concrete, and by the next morning all the sand would be blown out and we'd have to start over. The brick masons," he says, "would start out on ground level. The wind would begin to pick up later in the day, and they would get up higher and higher on ladders, sometimes six or eight feet, then they would generally be high enough to get away from the real driving sand. We put on hard hats and goggles to go out into the plant because of the wind blowing so much through the tunnels we had built."

Hale continues. "We installed the indigo-dye range without even having a roof over the building. After we installed the range, and once we had the roof and the walls up for the building, we found out we had to go back and take all the bearings off the range and get all the sand out of them and repack them with grease before we could start up. We were operating looms at one end of the weave room when bulldozers were still working on the foundation at the other end. That sounds cockeyed, but by doing such drastic things, we had denim cloth to sell and some income coming in immediately.''

Pete Jacobi of LS&CO, who then was general merchandise manager for the Jeans Division and successor to Tom Kasten, recalls one of his first visits to the site. ''There were spinning frames and looms already in place. In some cases, they were operational, [yet] you could walk through an open hole in the wall and be out in the middle of the desert.''

''Of interest to me as an engineer,'' Hale says, ''is that we didn't have any air conditioning, any humidification, and everybody says you have to have at least 85 percent relative humidity to be able to weave cloth. All textile plants had about that amount of humidity in them. A normal weave room is just a complete fog . . . moisture on the floor and on the looms, [supposedly] so you could run. We started up and ran looms out here when the relative humidity outside was seven percent and inside it was five percent.'' The conditions were not ideal, he admits, but ''we could run. We didn't have any lighting to speak of, either. We only had a few incandescent bulbs hanging down. So another problem we had was 'bug juice.' Bugs would be flying all around the lights and would fall into the cloth and be rolled up. You'd get a 'squeeze' of the bug, and that bug juice was cause for [the fabric to become] seconds. It may seem a little silly [for us] to [have taken] those drastic measures, but we had to show we were going. It was of primary importance. And,'' he reiterates, ''we needed to have some cloth to sell.''

As the plant was being built, so was the workforce. Danny Davis recalls, ''We had begun to hire experienced supervisors, a lot of people from the eastern part of the country . . . Georgia, Alabama, South Carolina . . . the textile southeast. Probably by May, June, or July we had more of them than we had local people in supervision.'' He remembers the crew with which they had started, the ones Bob Hale called ''raunchy.'' Davis says, ''It didn't take too long for [the bad ones] to show. Most of the people we hired would come and stay a few months and leave. The better ones generally ended up staying because they could see that it was a good opportunity.'' At that time, he says, ''the local workforce was probably grossly substandard [compared with the local] workforce we have now.'' He attributes much of that to the fact that the plant "paid somewhere barely above minimum wage" when it started. "We had a fairly good economy at the time [and so] we didn't attract a lot of the top-notch people in the area. Looking back, it probably would have been better to start paying people a little bit more."

On July 4, 1976, *Cotton Clips*, the ACG newsletter, explained the meaning of the bicentennial of the United States to the member-owners of ACG. "People like Thomas Jefferson, himself a farmer, who drafted the Declaration of Independence, meant for our citizens to have no limits on their pursuit of good. Our country's greatest resource," it was written, "is the right to dream, venture, and contribute to our society. This right and the 200-year history of the adventure of self-government is the inspiration of American Cotton Growers." American Cotton Growers was a late-twentieth century institution that the framers of the Declaration of Independence would have understood, and with which they would have felt a strong bond.

Some ACG members plainly were tired of dreaming, venturing, and contributing. Problems involving the pool kept building. One of Doyce Middlebrook's neighbors was a pool member who defaulted on his marketing agreement by failing to deliver his cotton to the Shallowater Coop Gin. He had to be threatened with legal action. Middlebrook recalls, "It was very frustrating to me to be on this board and to have something like this come up in my district and at the gin where I ginned." To make matters worse, "the man lived across the street." Middlebrook explains. "[My neighbor] had always tended to gin his cotton at an independent gin. [ACG] looked good to him at the beginning, but during the course of the year he decided that he didn't want to put his cotton into the pool, and he didn't want to have to wait for his total amount of money. So he took it to another gin and ginned it and sold it to a cotton buyer in Lubbock. The thing that was interesting," he adds, "was that we examined the number of bales of cotton that he violated his contract with and it wasn't a large number of bales at all." The farmer settled with the association by paying $35 in damages for "nonperformance" on each bale of cotton he had not delivered to the pool.

Middlebrook was the youngest of the ACG board members. In his early thirties when elected to the board, he is a generation removed from the others. He is a friendly man, not only willing to talk but conversant in a wide range of topics. He and his wife believe deeply in community service. In addition to his cooperative activities, including 14 years on the ACG Board and several more as vice-chairman of PCCA, both husband and wife serve as volunteers in community activities. Middlebrook is both past and future in West Texas cooperative agriculture. He is squarely between generations of farmers. No age group has exclusive rights to any trait, but it is fair to say that Doyce Middlebrook represents certain characteristics generally associated with the younger farmers and others commonly attributed to their elders. He is progressive, open-minded, and well-informed. But he also is mature, thoughtful, and possesses a razor sharp sense of responsibility. "Responsibility" is a word he frequently uses, and so it is understandable that he expects others to abide by contracts and agreements.

It was not long after the first yards of denim rolled off the looms at Littlefield that the Lea County Gin finally announced it would have to fold. It simply could not sign up enough producers to make its participation worthwhile. It petitioned the board in July to release it from the textile pool marketing program, and the board reluctantly approved the action.

Still later in the year, a pool problem arose at State Line Cooperative Gin. A contracted farm had been sold to a buyer who had no intention of honoring the textile pool marketing agreement. The agreement, which was on the acreage rather than the producer farming it, would stand. Bill Blackledge was authorized by the board to notify buyer and seller that ACG expected them to perform on their contract or it would seek damages. Blackledge made clear ACG would not hesitate to take legal recourse to enforce the agreement.

On Wednesday, July 17, the first load of Littlefield denim was delivered to Levi Strauss & Co.'s Midland, Texas plant. And on July 29, ACG held its board meeting at the mill for the first time. Bob Hale took the directors on a tour of the new plant.

By October, he was able to report to the board that all 360 looms would be installed and operating by January 1. But he also reported a serious problem of labor turnover at the mill.

Dorothy Hill, today senior payroll clerk at the plant, started working there in October, 1976. She remembers the problem and how bad it was. At the time, she was training to be a weaver. "There were several of us who started on the same day. About 10 o'clock in the morning, this other girl said she left her lunch in her car and was going to get it. She never came back." Hill shakes her head, "Things like that were happening all the time back then.

They wouldn't last but two or three hours, a day or two.'' Not that Dorothy Hill is completely unsympathetic to the adjustment problems of others in the fledgling operation. When asked if she had reservations, she answers, ''Oh, yes, and would have been one of those who didn't make it the first three days, either. But my kids, my family, thought it was so funny that Mother was going to work because I had never worked. And I'm just stubborn enough that I was going to stick it out. And I'm still sticking it out.''

The turnover rate was 200 percent. Danny Davis says the problem was so bad that ''in the first year of operation, we had 750 people working here, and we had to go through about 1,500 people to come up with that.'' The high turnover and resulting personnel inexperience were major contributors to the quality control problem that tested and strained the skill of ACG's directors and managers, as well as the relationship with Levi Strauss & Co.

The "Denim Dream" Becomes a Nightmare

THE QUALITY BATTLE

Bob Hale says, "The battle to meet Levi specifications was the greatest battle that we had from the very beginning." Levi Strauss & Co. had contracted with ACG to take all the fabric the mill could produce. But there was always that one condition, the loophole you could drive a module truck through. The fabric had to meet LS&CO's specifications. And there are no more stringent specifications in the denim world. "I can see why the contract had to be written like that," Hale says, "but it was an absolutely open door to them. They didn't have to take a thing."

Werner Pels talks about LS&CO's quality control process. "We try to stipulate very objectively exactly what we want to see in terms of both appearance quality and performance quality." The two criteria – appearance and technical performance – are the bases on which Levi Strauss & Co. tests fabric samples. Waverly Watkins of LS&CO was responsible for appearance quality, which Pels describes as including ". . . the things that you can see and touch and smell. The things that are obvious when you look at a piece of fabric – whether it's the right feel, shade, shape. Whether it's long enough, wide enough, things like that." Pels himself was responsible for those ". . . things that are not apparent, that you have to test. Performance characteristics, such as durability, tensile strength, abrasion resistance, colorfastness characteristics."

LS&CO would send people to the plant to provide evaluations of the appearance quality, because "it was easy to make judgments on the scene and . . . communicate what our concerns were," Pels says. "The testing part of it was satisfied by having samples sent back here to our laboratories so that we could spend some time — two or three days — to do the tests that were necessary.'' Levi Strauss & Co.'s main laboratory in San Francisco is the most comprehensive of the firm's testing centers. It conducts initial evaluations on new materials. ''Once something is in production,'' Pels says, ''once it's accepted and subsequent submittals have shown that the shortcomings have been worked out, once it begins to be shipped to our plant, then we monitor conformance to quality by two other regional laboratories in the field at El Paso and Greensboro [North Carolina].''

Bob Hale says, ''Each truckload of goods we sent them had about 26,500 yards of denim. They would sample the fabric, take a sample out of every designated number of yards. Then they ran 99 tests on the sample. If *any* of the tests failed, they sent the entire load of cloth back.''

Hale and his crew kept designing and redesigning methods for improving the quality. ''We changed the number of strands of yarn in warp and filling; the number of strands compared to the size of it, compared to the twist in it, compared to the blend of cotton in it. We worked to bring it all up to physical standards — warp tear, filling tear, warp tensile, filling tensile, elongation, stiffness, warp shrinkage, filling shrinkage, all the physicals.

''Then we had to meet the visual quality standard. All our equipment was brand new. We made a piece of denim that was *too* clean. It was *too* good-looking. They said they couldn't accept it; it had to look worse, it had to look rough. . . . ''We said, 'We've done all this to make a beautiful piece of cloth and you don't want it to look this good?' They said it just didn't look enough like denim.'' Hale was concerned that if his mill produced denim that looked worse and could pass the appearance tests, it would not be strong enough to pass the physical quality tests.

The plant clearly was not making what Levi Strauss & Co. wanted. Truckload after truckload was rejected. To Hale, it seemed that LS&CO was shipping more cloth to Littlefield than Littlefield was shipping to them. He told them, ''I can't believe this! You must be buying it from somebody else and shipping it to us.'' As if the rejection itself wasn't problem enough, ACG had to pay freight both ways on the loads that were returned. The situation got so bad the plant's industrial engineer told Hale the dock area would have to be rebuilt because it wasn't designed for receiving goods. Hale did not want to entertain the idea.

Frustrated, he responded, "No, the idea is to ship goods out, not get them back."

Every now and then, even in these troubled times, something went right. While the mill was producing denim that looked "too good," the standards of the marketplace suddenly shifted. "Women began wearing denim more," says Hale. "They liked the cleaner, dressier look. Then men began to pick it up and they liked the look. We reached the point where Levi said, 'Your denim looks good. This is what we want now.'" That helped stop some of the rejections.

Another production problem the mill had in the beginning was that its cloth, made with 100 percent open-end yarn, had a "waffled" look when washed. "You wash a pair of ring-spun jeans for a period of time," Hale says, "and they look streaked. Ours look crinkled. Levi said we had to do something about that look." There wasn't anything ACG could do about it. But the marketplace made another of its unpredictable shifts. "Now it is interesting," Hale adds, "that effect is what is wanted in denim. Beginning a couple of years ago, our denim was preferred over everyone else's because of the after-wash appearance. Those things are just pure luck."

Levi Strauss & Co. was dispassionate in its evaluation of ACG's denim, but it was not that way at all in its efforts to help its supplier solve the quality problems. Hale speaks, for example, of Waverly Watkins. "I worked with Waverly from the beginning, and he has truly been our friend. He told the cutters at Levi, 'Look, these people are starting out. They're going to get it right, but we have to work with them. Let's try it.' By him working with the cutting plants, they would not reject the cloth because it was too stiff or had a wavy selvage. He was wonderful to us."

Another propitious circumstance for ACG at that time was a tremendous worldwide demand for denim. The plant could sell its off-quality goods at a higher price than LS&CO was paying for top quality. ACG shipped denim to Spain, Italy, and other countries. Hale says, "Levi politely accused me of making seconds on purpose because I could make more money on seconds than on firsts." The insinuation, even made in jest, bothered him. "We were absolutely struggling to make every yard first quality to satisfy Levi."

Danny Davis says the yarn was part of the quality problem. "What I saw first of all," he says, "was the quality of the yarn that we purchased was not anything near as good as what we have today. When you run the yarn and it breaks, [you have to stop the system] and somebody has to fix it. With a lot of breaks in the yarn, you have a lot of stops. . . . The quality of the yarn and the inexperience of the people, I think, were the two big problems we had to start with. Plus, cotton runs best at certain temperatures and humidities, and back then we didn't have temperature or humidity controls. Now we run probably 60 to 70 percent relative humidity in the weave room. There are not many days in West Texas when the humidity is that high; and with low humidity the yarn was more brittle, and it broke more often."

As for the inexperience of the people, Davis says, "It took a while to develop that degree of skill. I think we could take the same people we have here now and go someplace and start up a plant and it would be much smoother because we have experience, and that's the biggest single thing." A statistic reinforces his point. "In the early days, we hired 70 of every 100 people who walked in the door. Now, of every 100 who come in and apply for a job, we hire eight or 10. We are much more selective."

Not only were the personnel inexperienced, the plant had its share of social misfits. Davis relates the story of an employee whose supervisor went to the canteen to tell him that he had taken too long a break. He responded by punching the supervisor in the face. Davis himself was threatened by an employee who took his job a little too seriously. The employee felt the process in the weave room was not being conducted properly, and he complained to Davis. Davis looked into the matter but did not take action that suited the irate worker, who told Davis that he would be "dead" if he ever caught him outside the plant. The employee was told to go home. About 15 minutes later, a report came in of a man with a gun in the plant parking lot. The logical assumption was that the man had returned to make good on his threat. It happened the report was false, but Davis says, "I looked over my shoulder for several days." The man who made the threat against him was discharged later.

An employee actually was murdered in the plant parking lot, but that did not happen until January, 1987, more than 10 years after the facility was opened and operating. It was the only severe act of violence in the mill's history. The slaying never was solved, but it had all the earmarks of a well-planned attack to settle a score. Other than the site chosen by the assailant, the incident seems to have had no relationship to the mill at all. It was a throwback to the old days on the Llano Estacado, the dying done on pavement now instead of dirt.

A less menacing incident from the plant's early days involved an unusually high number of yarn breaks in a particular employee's work area. "Every time the yarn broke," Davis noted, "the man was right there." Finally, he was caught in the act. "The guy was breaking the yarn, because if he broke the yarn the machine would stop more often, and he didn't have to put up as much yarn to run." He, too, was fired.

Davis emphasizes that these anecdotal incidents were much more the exception than the rule. Most of the employees, even in the early days, were stable, even if they weren't too skilled or reliable. The incidents simply reflect the dangers inherent in hiring large numbers of personnel in a short time. They probably could be recounted in every major manufacturing plant in America.

By November, 1976, there were significant improvements in the quality of the fabric produced at the mill. The mill, and ACG with it, appeared to be emerging from a quagmire of start-up problems.

BATTLE WITH THE BANK

The emergence would come slowly. The beginning of 1977 was not a happy period at ACG. The cooperative was going through some very difficult times. The denim market had fallen. In C. L. Boggs's words, "it was 'the pits.' Our quality was not what it should have been or later became. And I guess we had advanced 'even the loan' [the bare minimum covered by the government] on the cotton that year. We were seeing a lot of unrest on the part of our members."

On January 6, there was talk in the board of making a progress payment on pool cotton during the month of February. Although things were not going well, perhaps *because* they were not going well, the board felt it was crucial to have the farmer-members continue to see some return on their investment. "People outside the program were selling their cotton at higher prices," Boggs says, "and we had a lot of pressure from our members to make progress payments. We didn't have the money, and we didn't have much net worth, and the bank had to approve all progress payments and make the money available."

Don Scott of the Bank for Cooperatives explains how the progress payments work. "The initial advance [to the pool member] is exactly the Commodity Credit Corporation loan," what C. L.

Boggs called "even the loan." Scott continues. "Say the loan is 30 cents per pound, and the market price of the cotton is 35 cents per pound. The member of the pool gets 30 cents and the producer who is not a member gets 35 cents. So right from the start, the pool member gets less. . . . If the cooperative doesn't make money, then the pool member gets *only* the 30 cents per pound. If it does make money, he gets progress payments, which the cooperative hopes at the end will provide him a higher return than the non-pool producer. That year, ACG didn't make money, but they still wanted to make progress payments." Pool members, Scott says, "were watching their neighbors pocket $30 per bale more than they were. They could feel pain in their hip pocket."

Boggs says, "We didn't have any written contract with our members as to the regularity of payments or when they would be paid, but we had kind of fallen into a pattern of advancing even the loan, or a very small amount above it, then making the first progress payment in late January or February, and a second one perhaps in April or May. When you get into a pattern" he says, "people come to expect it."

ACG wanted the bank's permission to make three payments: a 10 cents per pound progress payment, another payment equal to the retirement of 10 percent of the building bonds, and a cash distribution of earnings from the 1975 crop. The board directed Dan Davis to broach the subject in meetings with representatives of HBC and the Central Bank for Cooperatives.

On January 25, Dan Davis called a meeting of the board to report on the meetings. The bank representatives had not been swayed. They were concerned about the cost overruns at the plant. And they wanted to see final arrangements completed for the loan guarantee from FmHA. Despite the bank's misgivings, the board voted to recommend that payments of about $13.2 million be made in February. These would include $0.6 million for retirement of principal on gin and regional building bonds; $12 million as a progress payment of 10 cents per pound on 240,000 bales of 1976–1977 cotton; and another $0.6 million in distribution of textile mill margins from processing 1975–1976 crop cotton during the last three months of 1976. In reality, the vote was nothing more than a plea for permission to make the payment. The board could pay nothing without the bank's approval. And that would prove very hard to get.

At a board meeting on February 3, Davis announced HBC had approved retirement of 10 percent of the building bonds. But it would approve a progress payment of only five cents per pound. And it deferred approving payment of the 1975–1976 outturn from the denim plant. The news hit hard. The directors were vocal in their surprise and disappointment at the bank's position.

HBC representatives frequently attended ACG Board meetings. Several of them were at this one and explained their position. The cost overrun was a major concern and would have to be funded from retentions from the current year's earnings. Further, there would have to be additional equity held by the association's members to ensure satisfactory progress toward the 1981 equity goal that had been established for the cooperative. That goal was one dollar in equity for every dollar in debt.

Though the mill had begun to turn the corner, it still was too far away from achieving quality to make anyone comfortable. Don Scott recalls, "We were concerned about the buildup of off-quality inventories at the mill. We just didn't know when they were going to get the problems worked out so they could make denim that Levi would take. And they were so highly leveraged at that time. We didn't feel like we could approve [the full progress payment] until they worked their problems out." The numbers painfully constructed by the ACG staff did not sway the bankers. "We went over their financial statement, we went over the projections," Scott says, "and at the time, with all the problems they were having, we just couldn't see. . . ." His voice trails off.

Three days later, bank and ACG officials would come together at the HBC annual stockholders' meeting, held that year in San Antonio. The ACG Board and top management had requested meetings with the full HBC loan committee to plead their case. People tend to erase unpleasantness from their memories. There were a number of meetings in several locations along San Antonio's Riverwalk, and to many of the participants they are just a blur. Boggs says, "I believe the full ACG Board was there. I don't believe Dan [Davis] and I attended a single session of the conference because we were in session with different bankers and with our board at different times."

ACG directors and staff used the opportunity to press hard for the 10-cents-per-pound payment. None of them was anxious to face the members if they failed to change the bank's position. The early morning of February 6, 1977, found them in the Menger Hotel in San Antonio, preparing for an 11:00 a.m. meeting with HBC officials. They studied files relative to the term loan from HBC. Projections of May 28, 1976, were reviewed and compared with new projections dated January 25, 1977. The staff reported on various ways in which funds to cover the cost overrun could be provided from estimated margins for fiscal years 1976 and 1977. A proposal to the HBC, including a required board resolution, was drafted for the meeting with the bankers.

Don Scott remembers the meeting. It was tense. In looking back on it, he simply says, ". . . I would rather not sit through another one like that." ACG was determined to get what it wanted and felt it should have. In this and subsequent meetings, the directors and staff argued their best case. But the bankers were immovable. They made a proposal for ACG to consider. The directors discussed it, and several changes were suggested, but in the end they were not satisfied it could be changed enough to help them. They would adopt the proposal, but only if the bank would approve the 10-cents-per-pound progress payment. It was to no avail. Boggs says, "They just turned us down."

L. C. Unfred feels the bank simply didn't understand ACG's situation, especially the people with whom they were dealing. It was important, he says, that the bank "stay with us. We had to do something for our members. But they didn't even want to negotiate on the progress payment." Unfred cannot recall how many times he and the other directors met with the bankers during those several days, but he says, "We almost wore out our shoes on the Riverwalk going between meetings."

The ACG Board reconvened the following day with a different purpose. Realizing now that the bank would not be swayed, the directors reviewed points to be covered at the gin community meetings scheduled later in the week. It was time to figure out how to break the news to the members. Dan Davis planned a slide presentation for the meetings, with information to be presented and questions answered by staff.

On February 17, the board met by telephone to hear a draft resolution that would tell the members that the association was required to hold their money to cover the cost overrun. "We will perform by September 30, 1977," it said, "our obligations to provide sufficient equity investment into ACG to cover during the current fiscal year the amount by which the cost of the textile mill exceeded the budget, including $2,000,000 in working capital, and to insure satisfactory progress in the formation of equity capital to meet the target of any equity-to-term-debt ratio of at least one-to-one by September 30, 1981." The resolution was passed, and a cover letter was drafted. The information would be mailed to each ACG member with the five-cents-per-pound progress payment that the bank had approved.

While the board and staff agonized over the bank's refusal to authorize the full progress payment, there was good news from the mill. In March, Bob Hale reported the Levi Strauss & Co. staff had been in "constant contact" with plant management, and there were no rejections of denim during January and February. He said gross sales during February would meet projections for the mill, although the number of yards produced was slightly under the estimate. This was only 10 months after the first yard of denim had come off the looms.

On April 1, the long-awaited loan guarantee was formalized in document-signing ceremonies. J. Lynn Futch represented FmHA, and his remarks were more hopeful than realistic. "This is one of the most stable projects we have ever supported," he said. The total loan had come to $33,500,000, and so the 90 percent guarantee was for $30,150,000. HBC representative David Dabbs presented Futch with a check for $301,500 to cover the one percent fee charged by FmHA for providing the guarantee. If the mill were to fail, America's taxpayers, through FmHA, would hold a bill from the bank for over $30 million.

A SOFTENING IN THE MARKET

The early progress of American Cotton Growers often seemed a case of "one step forward, two steps back." No sooner had the mill enjoyed two consecutive months without rejections, no sooner had the ink dried on the FmHA guarantee, than the denim market collapsed. The first indication for ACG was in an April 7, 1977, board meeting, when Bob Hale reported on softening of the market for denim in Europe. Less than a month later, and only one year after the mill produced its first yard of denim, C. L. Boggs told the board that even the market for off-quality goods was softening. For a firm that was trying hard to unload denim Levi Strauss & Co. wouldn't take, most of it in foreign markets, this news was distressing. The situation would get much worse. For the next year and a half, ACG would work desperately to find markets for its production and to manage an inventory that grew to monstrous proportions.

Tom Kasten of LS&CO says the softening in the denim market was exacerbated by contrast with the strong demand that had characterized the previous several years. "Supply and demand," he says, "got into better balance, and all of a sudden we found we didn't need as much denim as we were buying. We had to scale back our purchases significantly. American Cotton Growers was one [supplier] that was affected. We were their only customer and when we scaled back our needs, they got capacity they didn't know what to do with.''

Because of their special relationship, ACG fared less poorly than some of LS&CO's other suppliers. But the cooperative still suffered. It was too new and not yet successful enough to handle such a major adjustment. Kasten recalls, ''It was an issue of whether they could continue to make their payments to the bank against the debt. They were in big trouble. They didn't know what to do, and they looked to us for guidance. 'Should we develop a marketing department and start selling to other people? What should we do with all the denim here?' were questions they asked.

''It was pretty clear that the best option all around was for us to continue to support them.'' He emphasizes a point that is the foundation of the relationship between the two organizations. ''I felt we had an ethical commitment to them. If it wasn't for us, they wouldn't have made that investment, perhaps not built that mill and hired all those people. We had an ethical and moral commitment, as well as a business commitment to help them through this difficult period.'' LS&CO took as much of the denim as it possibly could. ''We took a bunch of yardage from them,'' Kasten says, ''and we often turned it into our own inventory. But we felt it important to honor 'the handshake' we had with them.''

It did not make ACG's competitors happy. Doyce Middlebrook says, ''We were the 'new kid on the block' [in the textile industry]. We were vulnerable at this time because we were weak. And yet we had this advantage by having the Levi Strauss contract and by making the commitment that we made in building this denim mill. . . . It was a tough time in the textile industry. . . . It was not just the pits for us, it was the pits for people like Burlington and Pepperell and Cone and Stevens, and all these other people. It was a tough time, and they didn't like the fact that they were having to cut back on their production of denim for Levi and we were still able to run six and seven days a week.''

Tom Kasten says that nothing in particular regarding the competitors' attitude toward ACG stands out in his mind. He does not remember much about the ''grousing'' to which Gil Melbardis refers. But he said he always found it interesting that ''some people never referred to ACG by name. They always said 'the Texas people,' or 'the cotton farmers.'

"When ACG was getting started, many in the industry were certain they could not last." The competitors, according to Kasten, probably thought "They don't have the skills. They don't have the expertise. Even if they have the money, it takes more than just money to produce a yard of denim that will meet Levi Strauss & Co. quality standards."

Once (ACG) was up and running, he adds, "I think they were accorded existence. . . . But nobody that I recall bad-mouthed them or ran them down or called us fools, to our faces anyway. Who knows what they were thinking? But these people are professional business people and they're not going to say anything that's going to have an adverse effect on their relationship with us."

Levi Strauss & Co. maintained an honorable position toward ACG. But even though they took all the ACG denim they could, it was not enough to keep the inventory from becoming a problem. How bad was the inventory problem? Bob Hale remembers it well. "At one time, we got up to nine million yards of inventory. We had denim stacked in the warehouse; we had denim stacked in open spaces throughout the plant, from one end to the other. We rented all of the cotton warehouses that we could find in the city of Littlefield.

"It so happened that West Texas Enterprises, which used to be a compress, had old cotton warehouses and we rented four of those. We had to reactivate the sprinkler system and put the warehouses back in operating condition. We had to put in lighting and loading ramps, put up a guardhouse, hire guards, and so forth." Hale says the inventory was "so tremendous you simply could not keep it dust-free and clean. Of course, we tried to take care of the cloth as best we could. We tried to wrap it up, and it stayed for a good while." Dust and dirt weren't the only problems. "We had a real problem with rats, because the cloth has both starch and tallow [a sizing material made from animal fats] on it. We put out rat traps and rat poison. Rattlesnakes love to crawl up into the rolls of cloth. It's a nice, shady place to get into. And then we had some people who tore the corrugated iron off the side of a building and stole some of the cloth." Had the mill been located in the humid Southeast, mildew would have been another major problem. Hale says, "Thank goodness it never rained enough that we had a mildew problem. But our cloth dried out. We had to regrade and reprocess three million yards of it so that it would look like new."

As the inventory kept building, the plant kept producing. "If we cut back operations as much as we might have," explains Hale, "we almost would have been shut down. And just to be getting on our feet and then shut down for week after week would probably have meant the end of operations. We had to keep operating and hope that the denim market would break and we could sell it to Levi or someone else."

The board still was trying to find a way to make another progress payment. A new projected statement of operations and margins was prepared on March 25. The projections were slightly better than those given the HBC at the San Antonio meetings. They would be sent to the cooperative banks with a request for approval of another five-cents-per-pound progress payment and payment of the remaining margins on 1975 crop operations. Months went by without the issue being resolved.

On June 2, 1977, the board appealed to the bank's representatives for approval to pay nearly $600,000 of the remaining 1975 margins. The bank told ACG management on the following day that the request would be looked at on a "month-by-month basis in trying to arrive at a time when the approval could be made." Three months after the meetings in San Antonio, there was a faint glimmer of hope in a dark desert of frustration.

Two weeks later, ACG got its answer in a letter from Andy Gilliard, vice-president of HBC. "The Bank's loan committee, in analyzing the overall situation," he wrote, ". . . reluctantly agreed to allow the $590,000 payment, provided that the American Cotton Growers Board and management would be willing to notify its membership that no further consideration would be given to pool settlements until sometime in September. . . ." At last, good news. But the letter continued. "This being a prior approval loan, [meaning that HBC could not grant approval without permission from higher authorities] our findings were submitted to Farm Credit Administration for their concurrence. . . . Farm Credit Administration did not concur with this approval." *Crash* ! The letter ended, "It was the Farm Credit Administration's opinion that the payment be withheld until the Bank knows with some reasonable assurance the results of this fiscal year and the true financial position of American Cotton Growers."

Then, in a display of empathy that must have seemed very hollow to ACG at the moment, the letter added that the "Farm Credit Administration recognizes the basis for this request and the contingency of adverse membership relations involved in not approving this payment.'' ''Adverse membership relations'' was a euphemism for ''there'll be hell to pay.'' It would be more than three painful months before the bank even would consider another payment to the growers.

As the fearful reality of the bank's response was sinking in, the board was presented a report by Bob Hale. It was from Levi Strauss & Co., and it said that ACG was rated their ''number one supplier for quality . . . for the preceding quarter.'' After one year in operation, the mill had achieved a landmark. It was the best of the best.

In retrospect, the achievement was not much more than a fluke. It was less the result of ACG rising magnificently above the crowd than of LS&CO's other suppliers all falling down at the same time. And coming at a time when problems at the mill were rampant, it almost seemed a cruel joke. It was an achievement that would not be repeated for another four and one-half years. But at least all the questions about whether LS&CO-quality denim could be produced from West Texas cotton using open-end spinning seemed to have been put to rest. And ACG would take any good news it could get.

The mill still was trying hard to unload inventory. The domestic market was hopeless. The foreign market was not much better. Hale told the board an initial order had been secured from a Danish cooperative, and he was optimistic about possibly securing the business of Swedish and Norwegian cooperatives. But this was grasping at straws. ''When the slump came in the denim market,'' Hale says, ''the slump came everywhere. You couldn't sell denim anywhere.''

The denim glut was a desperate situation, and desperation causes people to make mistakes. One of the sales ACG made in this period is one the cooperative would rather forget. According to Hale, ''We let a shipment of denim get out of the port of Houston, on its way to Spain, before we got our money or had a letter of credit opened. You just don't do that.'' Hale says he and some other people did not know it had happened, but others did. ''They put their trust in a fellow, and they shouldn't have.'' They found out later ''he wasn't good for the debt . . . that's all there was to it. The shipment never should have left.'' The man was a broker from El Paso; he worked for a firm called Longhorn Textiles. ''He had paid all right in the past, and this was another shipment he bought,'' Hale says. ''About the time the ship was in mid-ocean, the Spanish government passed an embargo on imports of denim from the United States. Well, here we were with a shipload of denim heading for Spain, and they wouldn't let it be unloaded when it got there.''

The ship was diverted to Rotterdam, and the denim was put in a warehouse there. ''Talk about warehousing problems,'' Hale says. ''The warehouse was down in the dock area. A lot of the denim mildewed; it just rotted. We tried to sell it from there. We tried to cover it with insurance, but the fact that it was getting wet and rotting made it hard to get insurance. We finally sold it for whatever we could get for it. A large portion of it was finally shipped to Italy, and part of it was shipped back across the Atlantic, to Manhattan Textiles in New York.'' Hale went to see it, ''because a man called me and said the stuff was rotten.'' Hale thought the man was complaining about the quality of the denim. ''I told him I knew it was not good quality, but to say it was 'rotten . . . .' He said,'No, I mean it's *rotten* !' I went up there, and they had container loads of it on the dock by the East River. I remember I almost fell into the river getting the first container open. And when I finally did, it smelled worse than any outhouse you have ever smelled.'' When the man said ''rotten,'' he wasn't exaggerating. ''It was so rotten,'' says Hale, ''You could reach in and pull out handfuls of the cloth.'' Hale calls the entire episode ''probably the worst fiasco [the mill] ever had. It was just a total calamity.'' Though he was not responsible, he would prefer not to have his name associated with it. Smiling, he says, ''We don't talk about that around here.''

In the July, 1977, ACG board meeting, the Commodities Futures Trading Commission case against PCCA, then six years old, was on the agenda. The government had proposed a settlement. Though the case was against PCCA, there were overlapping effects and relationships the ACG Board had to consider. The directors discussed the advantages and disadvantages of continuing litigation versus reaching the settlement proposed by the government, a one-year suspension from trading for PCCA and Dan Davis.

The PCCA Executive Committee had recommended that a settlement was preferable to continuing litigation. The ACG Board voted its concurrence. Then, to put distance between their association and PCCA, at least in the matter of futures trading, the board voted to modify their agreement with PCCA. The modification made clear that all futures marketing and hedging operations of ACG would be the responsibility of that association and its personnel, and that "PCCA and its personnel would not have any direction or control over those operations." "Personnel" in this case clearly referred to Dan Davis. At a subsequent joint meeting of the ACG Board and PCCA Executive Committee, the policy was reviewed and a proposal made that ACG handle its own futures market trading by employing C. L. Boggs, Davis's bright, aggressive assistant, as its general manager *and* futures trading manager.

In August, the proposal to employ Boggs as general manager was taken up by the ACG Board. Billy Whorton moved that it be approved. The motion was seconded by Doyce Middlebrook and passed unanimously. C. L. Boggs had ascended to the top management position at American Cotton Growers. No longer a caretaker of ACG for Dan Davis, Boggs now had full management responsibility for the cooperative and its future. A native of Quitaque, Texas, Boggs joined the PCCA staff as office manager in 1965 and became assistant general manager in the following year. Even before moving into cooperative management, he had served the cooperative cotton industry in Texas for nine years as a certified public accountant.

In a September 23 telephone meeting of the board, the new general manager advised the directors that HBC had responded to their latest request, this one for a four-cents-per-pound progress payment. David Dabbs, an HBC vice-president, said the bank had considered a two-cents-per-pound progress payment but upon further review that one cent per pound was the maximum progress payment that the cooperative banks would permit ACG to make at that time.

At a meeting the next day, ACG Chairman Unfred talked with Dabbs about the problems facing ACG as a result of the bank's refusal to approve the higher payment. There was nothing Dabbs or anyone else at HBC could do. He said the very small amount of equity capital from ACG members caused the Farm Credit Board to insist that the cost overrun on the plant be withheld not later than September 30 as equity capital from the 1976 crop or other sources.

Three days later, Unfred conveyed this information to the board. He told them he had been unsuccessful in persuading the bank. One cent per pound was the limit. The board immediately voted to make the payment prior to the end of the ACG fiscal year on September 30. Even a small payment was better than nothing at all. In addition to the progress payment report, the board reviewed proposals for handling the increase in membership relations problems which were expected. The directors reviewed a schedule of proposed pool gin membership meetings, a proposed script of a slide presentation for the meetings, and a letter to the banks and financial institutions notifying them of meetings to review ACG operations. The gin meetings are described by Dan Davis today as having been "pretty traumatic."

The staff still was living with the inventory problem and searching for outlets. At year's end, 88 percent of the mill's production was meeting Levi Strauss & Co.'s specifications. The problem was not one of whether the fabric was good, but whether it could be sold.

Denim was in trouble at the start of 1978. So was cotton. In fact, the general state of agricultural America was poor. An organization called the American Agriculture Movement had been formed to call public attention to the plight of the family farm. It was a national response to a national crisis. Farmers saw their situation as partly due to Carter administration agricultural policies, mainly the embargo on grain shipments to the USSR in the wake of the Soviet invasion of Afghanistan. They felt the embargo was more punishing to them than to the Soviets. The Carter policies, the state of the economy, and the difference in the two parties' approach to national security and "traditional values" were factors that helped put American farmers in the Republican column in the 1980 presidential election and kept them there in 1984 and 1988, as well.

ACG took its place beside the nation's other farmers in the American Agriculture Movement. Bob Hale told the board that businesses in Littlefield had agreed to close from 9:00 a.m. to 6:00 p.m. on January 9, 1978, in support of the movement. The board approved a motion by Doyce Middlebrook and seconded by Harold Barrett that ACG close its office in Littlefield and stop shipments for a 12-hour period in a gesture of unity.

It had been a year since the confrontation on the Riverwalk, and representatives of the Houston Bank for Cooperatives were making it clear that there was no chance at all of ACG being allowed to make additional progress payments, at least in the foreseeable future. Nor would the bank approve the retirement of building bonds that had been proposed.

In late March, L. C. Unfred reported on meetings where he and Dan Davis had attempted to obtain a commitment from the bank that no capital would be retained from lint proceeds from future crops. Predictably at this point, the bank was unwilling to make that commitment. The reaction to the bank's position was evidenced by mass sign-outs of acreage from the pool, a steady flow of members from ACG that lasted through the year. The loss of members would jeopardize the existence of ACG more than any other peril it faced.

The ACG staff discovered it had made a bad situation worse by carrying the huge denim inventory at cost of production rather than market value. The net effect of that was to make the financial picture of the association look even worse to the bank than shown on its statements. And that was bad enough. The staff reevaluated the denim inventory and wrote it down in value by $4.5 million. The write-down, although necessary to reflect the realities of the cooperative's financial picture, further heightened the concern of the bank. It meant a major asset of the cooperative was worth far less than thought. Doyce Middlebrook says the whole situation was "just like knocking down dominoes. Everything was coming apart at the same time."

In late January and early February, the board and staff focused on two topics: what it would tell the members to get them to stay in the pool, and how the pool policy might be amended to make it more realistic and attractive. As for what to tell the members, they carefully prepared for upcoming pool membership meetings. A script was written, and it was reviewed and revised numerous times. One of the tools that would be used in the meetings was a videotape prepared by the staff and featuring remarks by Tom Kasten of LS&CO. Kasten, then general merchandising manager, was the buyer for all denim his company used. His remarks were aimed at bolstering confidence in the future, confidence that the denim market would once again be strong and that financial success for the mill and its owners might yet be a reality.

Kasten says "We had been through ups and downs in the denim cycle before. If the market was heading into a trough, we knew it would turn around and start to head up toward another peak. We just had to get through that." Of LS&CO's decision to buy from ACG quantities of denim it did not need during this difficult period, Kasten says, "It was a very difficult decision because it was *not* in the best interest of Levi Strauss & Co. in the short term. We were a public company then. Every quarter we met with analysts, and this was definitely going to impact our inventories." But he quickly adds the decision was "clearly in our interest in the long term. [LS&CO] is a very ethical company. It has high values, high standards. We've spent 135 years building on our values and our integrity, and that's what we are known for. [Taking inventory from ACG at that time] may not have been what Wall Street would have wanted to see done,'' he says, ''but it was consistent with how we want to run this company.'' No one knowing Levi Strauss & Co. would be surprised by its unorthodox, ''unbusinesslike'' response to ACG's plight. The firm is living, growing proof that financial success and social responsibility not only are compatible, they are complementary.

Peter Haas, Sr., former chairman of the board of LS&CO and great-grandnephew of the founder, traces its corporate philosophy, one built as much on durable, reliable relationships as on durable, reliable clothing. ''I suppose the whole thing goes back to my grandparents and my parents and their involvement in the community and feeling that we were fortunate to have what we had,'' he says, ''and there was an obligation to give it back in time or money to the community.'' The concept, prominent in the teachings of the Hebrew scholars and clerics, is one the family not only has embraced but brought into its corporation. It is manifest in contributions of time and money to a wide range of charitable and civic causes, including a far higher percentage of revenues than is emanating from most other corporations. Just as important, it is manifest in business relationships. ''Usually,'' says Haas, ''we do a lot of these things because we think they are right . . . whether it's dealing ethically with business or with our own people.'' There seems never to have been any question that, having formed its relationship with ACG, Levi Strauss and Co. stood prepared to see it through.

Trouble in the Family I
The Pool Drains

The searing issue of pool membership somehow had to be resolved. The trickle of lost members was becoming a flood. A special subcommittee of the textile pool committee was elected in district caucuses to assist the ACG Board in developing a policy on "split ginning." Split ginning would not kill the 100 percent concept, but it would crack it wide open by permitting pool cotton to be ginned at non-pool gins and vice versa. The policy would not be entirely at the discretion of the producer or gin. Those gins that could stay with the 100 percent concept would be required to. The subcommittee members' names were announced early in March. They were Jim Bob Curry, Ernest Schattel, Wayne Huffaker, Wendell Norman, W. D. Vardeman, D. C. Cornelius, and Charles Edgemon.

Wayne Huffaker, a producer ginning at Farmers Coop Gin in Tahoka and who the following year would win a seat on the ACG Board, was a supporter of the 100 percent concept . . . until he saw the effect it was having. "As the financial difficulties came along, the local gins, especially our[s] . . . started losing a lot of membership because of it. . . . At the time we went into the pool, our local gin was ginning 52 percent of the cotton ginned in the local community. After about two or three years of ACG," he says, "we dropped down to 30 percent. We lost a lot of our volume because . . . those growers who were willing to participate weren't willing to put in 100 percent of their crops. They wanted to spread their risks out." Huffaker became an outspoken opponent of the 100 percent concept.

He says many ACG gins, through the representative each had on the textile pool committee, expressed adamant opposition to the concept by this time. It was partly because ACG had said several years earlier that it would not take any lint proceeds to finance the operation, and then had to do it to satisfy the bank when the denim mill lost money. And it was partly because growers outside the pool were receiving more cents per pound than those in the pool. But it also was because the gins and growers were under pressure.

"Their bankers," says Huffaker, "would tell them they could have part of their cotton in the pool but to keep some out to market without a risk. There was pressure from the banks, pressure from the landlords. If they were taking losses because of the denim mill, they told those growers not to participate. Even though a lot of them wanted to participate, a lot of them couldn't because of the 100 percent requirement. . . . Some growers would have two landlords, and one would say, 'Yeah, you can put my farm in the pool,' but the other one would say, 'Don't put mine in there.' It caused conflict where the grower was forced to drop out of the pool. . . ." When that happened, the gin lost a customer. "There were growers in our community," says Huffaker, "who were forced over to the independent side and still haven't come back."

Who was in favor of the concept? Huffaker maintains "It was the entire . . . board of directors, even though we were able to beat them in a straw vote in the pool committee. On several occasions when votes were taken, the results clearly showed the membership at large didn't want the 100 percent concept." Huffaker doesn't entirely fault the board for its position. "The board of directors continued to stay with that 100 percent because of pressure from the Bank for Coops. They were telling the board . . . 'You *will* be 100 percent or we're not going to finance you!'" Nonetheless, Huffaker's sympathy for the plight of the board is lukewarm. "There wasn't a whole lot they could do about it." Then he adds, "But I've often thought that there was *more* that they could have done." The approach he suggests would not surprise the people who know him. "They could have told that bank to go to hell." Asked if the bank would have backed down, the outspoken Huffaker replies, "They would have had to. They had $40 million lying over there."

Although willing to consider revision of the 100 percent concept, the board let it be known it was ironclad on its agreement with the producers. It adopted a resolution telling producers that they were either in the pool or out of the pool until such time as the board changed the official policy. Having reaffirmed its position, it directed the staff to review problems related to the concept of split ginning and to work with the special study committee to develop recommendations.

C. L. Boggs and the staff tried to find a middle ground. One extreme would have been to stay with the 100 percent concept, which seemed at this point suicide by bloodletting for the organization. The other was complete removal of restrictions on the gins, which the bank would not permit and which might create chaos and remove any incentive for producer and gin loyalty.

At a special study committee meeting on March 9, Boggs presented an option of staying with the 100 percent concept as a long-range goal, but permitting split ginning, also called "dual ginning," as a temporary alternative. It would be a way of helping those gins in the most severe financial trouble due to lost members by allowing them to serve both pool and non-pool producers. The key questions raised included how can we resolve problems that might develop among the 100 percent gins if others are allowed to dual gin? What are the incentives for gins to stay with the 100 percent concept if dual ginning is allowed at others? Would a new plan be more feasible if, after a member signed out, a waiting period of three years would have to pass before he could sign back in?

Throughout March, 1978, the ACG Board and the special study committee continued discussion of the dual ginning proposal. The directors wanted to consider in more depth the effect that the proposal, now called "Alternative A," might have on the pool gins prior to making their decision. For its part, the staff continued to prepare increasingly detailed analyses of the measure for the board to consider. The staff also talked with non-pool PCCA producers about the proposal. Among them were Jim Bob Porterfield and Ted Aten. The reaction of the PCCA producers was not exactly what the ACG Board wanted to hear. They suggested ACG develop a comprehensive plan for conversion to an "individual producer base." In other words, "let the farmer decide." With such a wide gulf between the positions of the two cooperatives, each largely dependent on the other, a summit was needed. A joint meeting of the ACG Board and PCCA Executive Committee would have to be held before ACG could reach a decision on its future plans.

ACG Director Joe Leach wanted to force the issue and pressed for a deadline on the decision. He proposed the agreement with the gins be amended to state that, unless a satisfactory alternative was adopted before February, 1979, those gins that could not continue to gin only for pool patrons would be required to withdraw from the program.

On May 1, 1978, the joint meeting took place. The agenda was weighty. An onerous discussion of the pool policy was scheduled, but there was to be a major surprise, as well. The staff reported the hemorrhage of members was getting worse. About 20 percent of the acreage in the program was signed out during February. At half the pool gins, the percentage of acreage signed out was not critical to future operations. At the other half, however, the sign-outs were significantly higher than average. ACG hardly could afford to lose half its gins.

The staff aired pool proposals being considered by the special seven-man study committee and the ACG Board. The meeting minutes reflect that "four plans had emerged," although the first three seem to be options and the fourth a future condition to be imposed on gins leaving the pool in whole or in part. The "plans" were: continue with the 100 percent pool gin concept, convert immediately to an individual producer base, convert to a combination of individual producer base at some gins and remaining 100 percent at others, establish a two or three-year grace period for the pool gins to return to the 100 percent concept.

After many discussions with PCCA producers and gins in the preceding weeks, the staff felt strongly about two issues. First, they said there would be resistance from some non-pool gins against any plan under which pool gins would be permitted to gin for both pool and non-pool members. Understandably, the non-pool gins thought it unfair if the pool gins were to have it both ways. Second, they felt the process by which ACG made its pool decision would have to be participatory. In order to be successful, it would have to include the opinions and receive the support of as broad a group of producers and gins as possible.

The textile pool committee had proposed that a 25-man study committee representing each pool gin be organized to consider solutions to the problem. It would replace the smaller, seven-member committee named in March. Staff recommended the proposal be approved by the ACG Board and any proposed solution be reviewed by the PCCA Board prior to adoption. After long discussion, the issue was set aside. Then, at this critical juncture on the first day of May in 1978, came a landmark in the history of both associations. Dan Davis announced his departure.

Turning Point

DAN DAVIS RESIGNS

After reporting on the status of cotton sales and the TELCOT network, Davis talked about his recent visits with major merchant subscribers to TELCOT. Based on his visits and market trends, he had concluded it was "inevitable that an electronic cotton marketing system would be made available to those who were not served by the cooperatives. "I thought it was important to the cooperative," he says, "that it be done in a manner that would bolster the cooperative program rather than be competitive with it." Ever the visionary, Davis decided to be the one to start that service.

The Agricultural Marketing Service (AMS) of USDA had a program called FSMET (Federal State Market Improvement Program) for which Congress routinely appropriated about $10 million, most of which was doled out in small grants of $10 thousand to $15 thousand each. The funds normally went to land-grant colleges to fund projects to improve marketing. The head of AMS had seen TELCOT in operation, was favorably impressed, and told Davis her agency was interested in making about $800,000 available as a grant to further electronic marketing. "I decided to resign," says Davis, "and start a company which would apply for that grant and use it as seed money to get started."

Davis would organize and manage a new corporation to provide electronic marketing services for the 60 to 65 percent of cotton producers who ginned at private gins. He did not want to sever his ties with the cooperative completely, but work out an arrangement to share hardware and software development and build a common buyer network. He suggested the PCCA Board consider joint operation of the data-processing equipment and leased lines by PCCA and his new company. He anticipated he would provide up to 300 terminals to private gins, and claimed that the volume marketed through a jointly operated system could lower significantly the per-bale cost of the PCCA data-processing operations. "In my opinion," he told the ACG Board and PCCA Executive Committee, "an electronic market open to all producers and buyers will generate better prices for farmers because it will stimulate competition. Competition will increase because small merchants will have market access equal to their larger competitors. And the expanded market network likely will attract additional firms to the handling of Texas cotton."

Davis tried to make the arrangement attractive to PCCA. His concept called for PCCA control of joint operations. PCCA's data-processing requirements would have first priority. His investment in computer equipment would bear a "reasonable relationship" to the service he provided to non-cooperative producers. His company would pay a "fair share" of the development cost of programs used in the venture. Operating costs would be divided on the basis of benefits received. And a procedure would be agreed upon in the event either party found the operation unacceptable and elected to withdraw. Davis said he had not completed budgets for the proposed venture, but if PCCA was interested, budgets and proposals for structuring the joint operation would be prepared.

The PCCA Executive Committee directed the staff to consider the proposal and develop specific recommendations. Two days later, the PCCA Board gave approval for a limited test on the 1978 crop to determine the feasibility of sharing computer and other electronic facilities with the new company, which Davis called Commodity Exchange Service Company (CXS).

As C. L. Boggs recalls, obtaining the PCCA Board's approval was not easy. "I think that's when I first understood that Dan had more detractors on our board than I realized, because when he made that proposal, there was a lot of opposition to it. There were two reasons. The primary opposition was that coop farmers who owned this business basically didn't want to share the technology with the independents who were their competitors. The second reason, I think, was that his detractors didn't want to share it with Dan." Boggs went ahead and pushed for the arrangement. "I sold it on the basis that it was a good business move and that someone was going to offer that service to the independents for sure, and then we'd have two systems, two overheads. I said we could lower the overhead for everybody if we did it this way. I sold it, and it continued to be controversial year after year after year."

Dan Davis never worked for ACG. Not "on paper." But from his position with PCCA, he was as instrumental in the development and management of ACG as any one person during the organization's first five years. Now his association with ACG would be over. To this day, he harbors misgivings about his departure. They stem from two causes, both related to timing. The first had to do with the business venture he launched. "In retrospect," he says, "my timing was very bad. The grant with which to launch the business did not materialize, and technology and the cost of delivering what was a very popular service was just too dang high. Computers were much more expensive in those days in comparison to what they could do, and lease lines were very expensive."

The contract under which PCCA and CXS shared hardware, software, and operating costs would last about six years. "I paid Plains some $2.5 million in that period that we shared the costs, but we didn't have any sources of revenue other than commissions from that trading operation. I was trying to convert it from the PCCA system to a PC [personal computer] dial-up arrangement, but we ran out of cash and went into Chapter 11 reorganization from which we had filed for bankruptcy in March of 1985. . . . We emerged from Chapter 11 last September [1987]. Plains set up TELMARK [a network for independents similar to TELCOT] and picked up about roughly half of the gins which our company had prior to our financial collapse. We're just really now getting back on our feet and getting started."

The second cause of his discontent was the circumstances of ACG at the time of his departure. In 1976, cost overruns and quality problems almost killed the venture. In 1977, it was the bank's withholding of progress payments and the sudden softening of the denim market. In 1978, it was mass sign-outs from the pool. After more than two years of ACG nightmares, one might think Davis's departure was hastened, or at least encouraged, by his farmer-employers. It was not.

Davis maintains that his decision to resign was his and his alone. And no one disputes it. C. L. Boggs says, "I can assure you there was no pressure at all for Dan to leave American Cotton Growers or Plains Cotton Coop, to my knowledge. I was his assistant, and I think I would have known if there were pressure. He made the decision to leave Plains and go into business for himself, offering TELCOT to the independents because he thought he was going to make a lot of money. That's why he left. There was a lot of regret that he was leaving on the part of the board," Boggs adds, "maybe not every board member, but the board as a whole. And there was regret that he was leaving on my part, because I didn't really want to manage American Cotton Growers or Plains Cotton Coop. I wasn't looking for the job, and I wasn't sure I was capable of managing either coop."

Asked if there was any sentiment from the board, staff, or any of his professional colleagues to get him to change his decision and stay on, Davis's answer is emphatic. "I would say overwhelming." Not that he doesn't recognize his detractors. He reaches inward for 22 years of PCCA history. "During the years that I managed the cooperative, we had a pretty aggressive and forward-moving posture. We were always reaching to try to get something accomplished and . . . there was always an element within the [PCCA] Board which was really against anything that we happened to propose."

Davis cites examples of controversies during his tenure, ". . . our consolidation with the cotton marketing associations in Oklahoma, the Rio Grande Valley, and Corpus Christi. Building the two compresses was controversial. The electronic marketing system was very controversial in its early days. Instrument testing . . . a lot of people thought that was the wrong thing to do. The consolidation of the gins at Crosbyton was very controversial. And certainly, there were those who were opposed to the denim mill and the creation of ACG. I would have to say," he adds, "that there were perhaps 10 percent of the [PCCA] board that, whatever we were for, they were going to be against. You knew that." Of the other 90 percent, he says, "I always had very strong support from the board. . . . The first year we lost money, with the 1968 crop, we lost four or five million dollars. Of course, I tendered my resignation to the board. But they would not accept it. The same thing happened a second time. We lost several million dollars on the 1971–1972 crop. I urged them to get someone else to run the thing, and they wouldn't accept that."

He talks about tendering his resignation as though it is something any decent manager would do automatically after a business downturn. As he admits, there was not much he could have done to prevent the losses. "We just advanced more for cotton than it later turned out we could get for it. In both instances, we made it all back and then some the next year. But cotton marketing is a tough operation. I don't know anybody engaged in [it] who doesn't have years that they lose money instead of making it.'' Dan Davis endured some of the worst times. By the time he resigned, the quality problems at the mill seemed to have been solved. Notwithstanding the loss of members, ACG was on the verge of turning around. At least it had not fallen completely apart yet. ''I would have to say, in retrospect, I probably resigned a year prematurely,'' he says. ''I think I had not done enough to prepare public opinion for the announcement. I think my supporters felt like I had left them when I shouldn't have and, I guess, my detractors were probably glad I was gone.''

On May 4, three days after Dan Davis announced his resignation, ACG held its annual meeting for 1978. The board and staff tried to put the best face on a grim situation. C. L. Boggs reported on two of three major problem areas: the slow progress being made in selling denim inventories, and the ''next'' progress payment, about which he could say only that he was unable to tell them when the next payment could be expected. L. C. Unfred reported on the third, the sign-outs from the pool and the deliberations of the study committee. Again, the only good news seemed to come from Littlefield. A month earlier, Bob Hale had reported the mill was producing 420,000 yards of denim per week, 88 percent of it first quality. At the annual meeting, he said there had been progress in reducing costs and problems in training and labor at the plant. Turnover was ''leveling off,'' and employee absenteeism had improved.

In the annual election of directors, the members of the association were not prepared to take out their frustrations on the board. All the incumbents had been renominated in their district caucuses, and the only one facing opposition was Doyce Middlebrook, who was challenged by Charles Edgemon for the seat from District I. All the incumbents were reelected. Immediately after the meeting, the board reelected the cooperative's officers by acclamation. It also approved the recommendation of the textile pool committee that the Special Study Committee of 25 ACG members, one from each of the pool gins, be created.

DEATH OF THE 100 PERCENT CONCEPT

The new committee got off to a quick start. It held its first meeting on May 9, just five days after being created, immediately electing officers. They were: L. C. Unfred as chairman, Wayne Huffaker as vice-chairman, and Bill Blackledge as secretary. One vote was authorized for each gin association. C. L. Boggs gave the members information developed by their predecessors on the seven-member study committee. Several members moved the ACG Board approve split ginning for the 1978 crop with the understanding that the committee would "work in harmony with the PCCA Board between now and next February 1 [1979] to develop alternative proposals which ACG might follow in succeeding years."

The quick start came to a quick stop. The motion to approve split ginning for the 1978 crop got no further. First, there were suggestions for amending it. Then the motion was set aside while the members discussed at length the purposes for which the committee was created. Finally, they agreed the committee would not approve anything so specific so soon. Instead, it would develop detailed plans for converting from the 100 percent pool gin concept "for further consideration and study." The motion to approve split ginning for the 1978 crop was withdrawn. Specific action was abandoned in lieu of more study. The 100 percent concept was still alive. Though his name is not mentioned in the minutes and no one recalls his identity, one intrepid soul moved that the committee consider the 100 percent pool gin concept dead and work during the next two weeks on developing a plan for an individual producer base. The measure likely was proposed by a representative of one of the gins suffering most from a loss of members. It was more than the committee was willing to consider, and it died for lack of a second.

Over the next few weeks, the committee met repeatedly and considered various permutations of "Alternative A," the dual ginning proposal, which by then had become "Plan A." "Plan B" had been developed, calling for immediate transition to an individual producer base. All the possibilities in that measure were analyzed and debated. The group inched closer to the only course it realistically could recommend. ACG could not survive with the 100 percent concept. But dual ginning was deemed unfair, unworkable, and unacceptable to the non-pool gins. The members voted paragraph by paragraph on a slow, deliberate conversion to Plan B. After all the language was approved, someone moved the complete plan be recommended to the ACG Board for consideration, subject to approval of the cooperative banks. The motion carried 16 to 4, with several committee members not present or not voting.

At the regular meeting of the ACG Board on June 4, C. L. Boggs presented Plan B to the directors. The board listened carefully, but took no action on the proposal for another month, when it was raised at back-to-back board meetings on July 5 and 6. Both Plans A and B were reviewed again. Joe Leach moved that the board adopt Plan B, and the pool be reopened for a two-week period during August. Dallas Brewer seconded the motion, and the vote was a deadlock, three for and three against. L. C. Unfred broke the tie by voting to kill the measure. They were back at dead center. Harold Barrett, a strong supporter of the 100 percent concept, then moved that Plan A be approved with a provision giving gins the option of deciding in February, 1979, whether to continue in the program. The motion was seconded by Doyce Middlebrook. Before taking any action, the board discussed the possibility of allowing each gin an option to reopen the sign-out period for its patrons. Barrett withdrew his motion. Now they were going in circles.

Then, finally, a solution. T. W. Stockton moved that parts of both Plans A and B be combined in a plan with the following provisions. (1) Each pool gin could decide by board or membership action whether or not it would gin cotton for non-pool members for the 1979 crop. This would bring enough business back to some of the pool gins to keep them from going under financially. (2) Each pool gin could decide whether or not to reopen the 1979 crop sign-out period for its members. This would provide producers wider latitude on whether to leave the pool or remain. (3) Non-pool gins would be permitted to gin cotton for pool members for the 1979 crop. This would make the non-pool gins happy, principally the members of PCCA who were not also in ACG. (4) The ACG Board would decide before February, 1979, whether to continue the 100 percent concept or convert to an individual producer base. (5) Each pool gin would be allowed to decide in February, 1979, whether or not to continue in the ACG program under the rules adopted by the board. The motion, later called "Plan E," was a perfect compromise. There was something in it for everybody — pool gins, non-pool gins, and producers. Billy Whorton seconded Stockton's motion. It carried.

The 100 percent concept was dead. The sign-out period would be set at September. Everyone would have until February 1 of the following year to reassess their position. Dallas Brewer explains why the 100 percent concept had to die. "Our local coop gins were losing so many customers to other coop gins who weren't in the pool. Even though we gin at different gins, we try to keep our coop gins going because we believe in the coop way of doing business. Our attitude," he says of when the board voted to change the policy, "was 'Let the non-pool farmers go ahead and gin at that pool gin, even though they don't want to be in the pool and textile mill.' It helped our gins financially." Even worse, as Brewer is quick to admit, they had been losing customers to the non-cooperative, independent gins.

Wayne Huffaker says a lot of the people who left the pool before the 100 percent concept was put to rest never came back. But ironically, after the concept was put to rest, most of the growers stayed in the pool. It seemed once the farmers were told they were no longer *required* to gin all their cotton at pool gins, many of them decided it was all right. "It was a big issue that put pressure on the local gins," he says, "and once that pressure was relieved, everything smoothed over." Asked if the real issue all along was coercion, he replies "It really was . . . and resentment against the bank and the ACG Board of Directors." It was that resentment that would put Wayne Huffaker on the board a year later.

George H. Mahon
(Photograph courtsey of Southwest Collection, Texas Tech University)

Roy B. Davis

Dan Davis

American Cotton Growers: Crosbyton Gin Division

L. C. Unfred, Chairman
1975-1989

Board of Directors, American Cotton Growers

Harold Barrett
1975-1979, and 1982-1989

Joe Leach
1975-1980

Dallas Brewer
1975-1989

T. W. Stockton
Vice-Chairman
1975-1989

Doyce Middlebrook
1975-1989

R. H. "Billy" Whorton
1975-1981

Board of Directors, American Cotton Growers

Wayne Huffaker
1979-1982

Clarence Althof
1981-1989

Jim Bob Curry
1980-1989

American Cotton Growers denim mill during construction
(1975–1976)

Denim mill in 1989

Bob Hale (left) and Dan Davis (right)

Levi Strauss and Company logo and slogan

C. L. Boggs

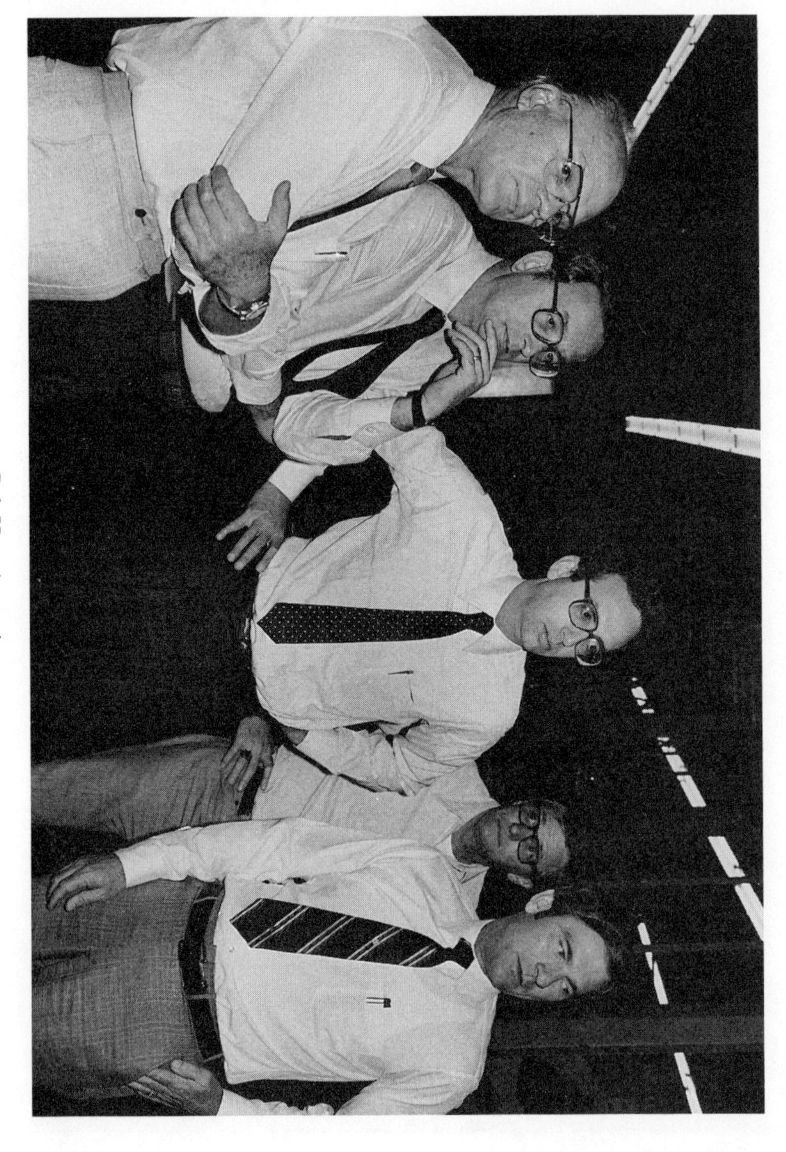

Bob Haas (center)

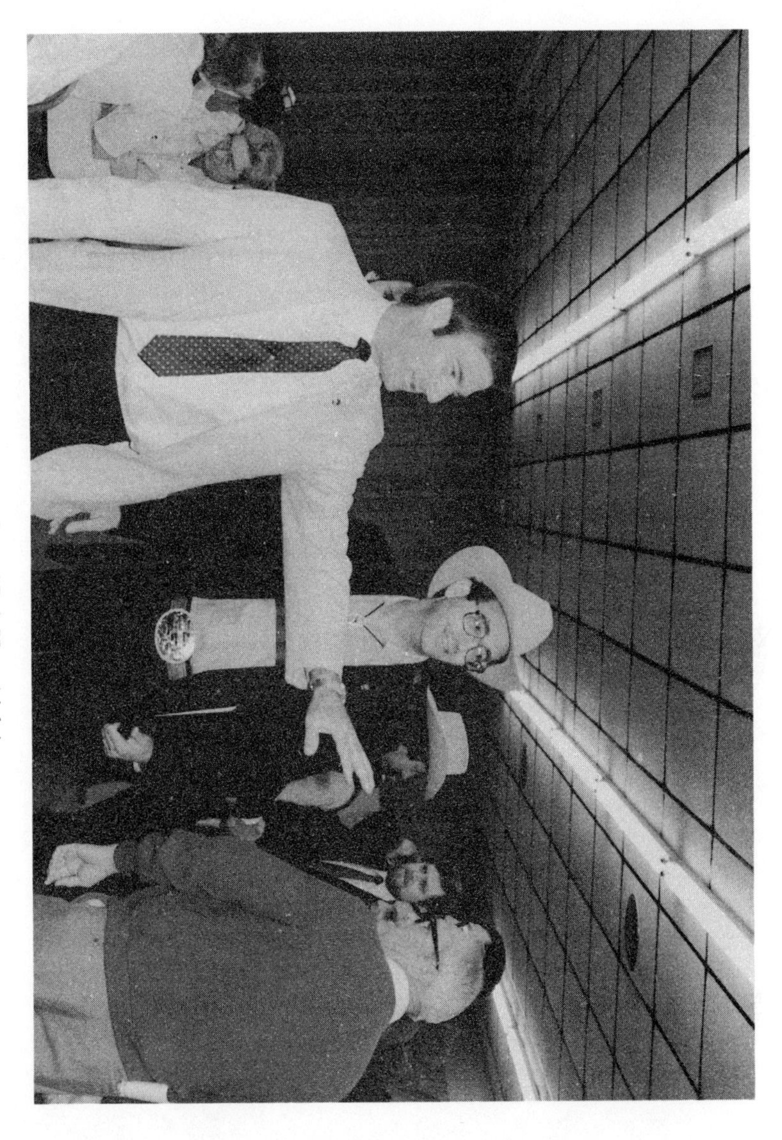

C. L. Boggs (left) and Bob Haas (right)

Present for the unveiling of the plaque were: *Row 1* (Kneeling or seated) left to right – C. L. Boggs, L. C. Unfred, Coy Grimes, John Johnson, Van May. *Row 2.* – Dallas Brewer, Harold Barrett, T. W. Stockton, Doyce Middlebrook, Jim Bob Curry, Harold Scaief, Wendell Jones, Tex Martin, Joe Rankin, Curtis Jensen, Charles Macha, Gary Sherrod. *Row 3.* – Bob Hale, Clarence Althof, Jackie Mull, Larry Lockwood, Eddie Smith.

AUGUST 2, 1987

A TRIBUTE TO PRODUCER-OWNERSHIP

This memorial commemorates a triumph of unusual scale by cotton producing members of American Cotton Growers and Plains Cotton Cooperative Association.

In 1975, ACG implemented an integrated concept of pooled raw cotton sales and textile processing. It built the first producer-owned textile plant, equipped to operate entirely on West Texas cotton, to manufacture 14 1/2 ounce, indigo dyed denim to meet the rigid quality standards of Levi Strauss & Co.

Over a 12-year period, ACG met many formidable obstacles but persisted and eventually achieved remarkable success that enhanced the economic level of its producer-owners and the many communities dependent on them.

As ACG achieved its unique concept, PCCA also shaped a new dimension in marketing with its innovative TELCOT system while serving ACG's needs in fiber testing and global raw cotton merchandising.

The quality and efficiency of ACG's denim mill attracted significant offers from the textile industry to purchase the facility,

In the search to maintain producer-ownership of this great asset, PCCA found the means to purchase the plant. Thus, the value added returns from the ACG concept became available to PCCA's large membership throughout Texas and Oklahoma.

This facility stands as a monument to the vision and deeds of agricultural leaders such as Mr. Roy B. Davis; Congressman George Mahon; Mr. Dan Davis, former general manager; and to many others who turned a dream into a reality. It could not have occurred without the kindred support of the producer-owned gins, regionals, cooperative banking institutions, their employees, and many others, including Levi Strauss & Co.

Listed here are producer-elected directors and officers who participated in the historic transfer of this facility from ACG to PCCA.

ACG BOARD

L.C. Unfred - Chairman
T.W. Stockton - Vice Chairman
Doyce Middlebrook
Jim Bob Curry
Dallas Brewer
Harold Barrett
Clarence Althof

PAST ACG DIRECTORS

R.H. Whorton
Joe Leach
Wayne Huffaker

PCCA BOARD

Jackie Mull - Chairman
Tex Martin - Vice Chairman
Larry Lockwood
Eddie Smith
Charles Macha
Gary Sherrod
Joe Rankin
Coy Grimes
Raymond Althof
Harold Scaief
Curtis Jensen

STAFF

C.L. Boggs - President
Bert Kyle - Vice President
Robert Hale - Vice President
Van May - Vice President
Darryl Lindsey - Vice President

In 1988, a plaque was installed in front of the denim mill commemorating the transfer of the mill from American Cotton Growers to Plains Cotton Cooperative Association.

STRENGTH PREMIUM RATES IN CENTS PER POUND

Crop Year	Grams Per Tex				
	24 & Less	25	26	27	28 & Higher
1982	0	1	2	3	3
1983	0	1	2	3	4
1984	0	1	2	3	4
1985	0	0	1	2	3
1986	0	0	1	2	3
* 1987	0	0	1	2	3
* 1988	0	0	1	2	3

*PCCA Pool

PREMIUMS PAID FOR STRENGTH

Crop Year	Total Premiums	Avg. Premium Per Bale
1982	$ 420,930	$ 4.43
1983	1,023,347	7.49
1984	1,989,797	9.46
1985	1,156,281	4.54
1986	997,868	4.80
* 1987	3,310,208	7.12
* 1988	3,932,752	5.18
	$12,831,183	

*PCCA Pool

Men of Achievement,
Years of Success

Men for Their Times

As was nearly predictable, C. L. Boggs was confirmed by the PCCA Board to be that association's general manager after Dan Davis's departure. It happened on July 5, 1978, and it was PCCA's first top management change since 1956 when Davis was appointed secretary-manager. It made Boggs only the third manager in PCCA's history. The first chief executive officer had been Ray Goss, who was hired on September 28, 1953, and remained in the position until March 13, 1956. Davis assumed his duties in June, 1956, and had resigned effective June 3, 1978. Boggs had been acting general manager during the search for candidates for the position, and it is hard to believe that there ever was serious consideration of awarding the post to someone else. He had, after all, been on the association's management staff for 14 years, 11 of them as assistant general manager under Davis.

In announcing the appointment, L. C. Unfred said, "Boggs has been an integral part of the cooperative management structure in the southwest for many years. He has been a part of PCCA during its most dynamic growth periods, including the development of TELCOT, and has assisted ACG from its outset." Boggs had been serving as general manager of ACG since September, 1977. He was placed in that position because Dan Davis *could* not be. Now he would be general manager of PCCA because Dan Davis *would* no longer be. Boggs was at the pinnacle of full-time management of both cooperatives. In that respect, he would complement Unfred, himself the president of both.

Boggs admits to having doubts about the new responsibilities being offered him. "I wasn't looking for the [PCCA] job and wasn't sure that I was capable of managing either coop," he says. "But when the job was offered to me, I felt like my options were to accept it, to stay here and work for someone else — and I didn't have any idea whom that might be — or to leave myself. I didn't want to do either of the latter. So I decided to take a shot at it."

Unfred said Boggs's dual capacity as general manager would mean greater efficiency and savings for cooperative marketing. "Though each association is a separate entity, there is a strong kinship between the two. Our goal is to improve the income position of cooperative producers whether they elect to market as a pool and in the form of denim or via the electronic service of TELCOT." In subsequent years, as competition between the cooperative systems intensified, Unfred would find that the double duty being done by him and Boggs would put them in one pressurized situation after another. But later still, it would lead to the solution of both cooperatives' most severe problems.

It is hard to believe that ACG or any organization could have two managers as different as Boggs and Davis, especially when one was tutored by the other. It is harder to believe that the organization could be well served by both. ACG was, but only because each manager was right for his time. Boggs, who worked for Davis for 14 years, describes the differences. Davis, he says, is creative, a "good planner." And using the term that has been used so often it seems part of Davis's name, Boggs calls him "a visionary." In contrast, Boggs calls himself a "hands-on kind of guy. I like to make things happen. If someone can lay out the options for me, I think I can analyze things pretty well and make a decision that more times than not will work." Boggs does not consider Dan Davis a good business manager. But he understands and appreciates the contributions Davis made to the cooperatives he served. "If I had been the manager of PCCA, going back to 1970, '69, '68, I doubt if Crosbyton and American Cotton Growers would have ever come into existence. I doubt if TELCOT would have ever become a reality, because I'm not sure that I would have had the vision to put all that together and make it happen." Davis had the vision, and Boggs helped him implement his ideas. When Davis left, Boggs was able to take what ACG and PCCA had at that time and make it work well. Boggs says, "I don't know that we've gone out and climbed any great new mountains since Dan left. I think that's the difference in management style. Is it right or wrong? I don't think one is necessarily right and the other wrong. It's simply a difference in style."

Boggs feels that Davis's innovative style tended to keep people off balance and created more discord at times than the organization could stand. "Dan's mode of operating kept things stirred up internally, because we had too many things going on at one time. We couldn't really zero in on anything long enough to do it right. [Being] five or 10 years ahead of our membership created a lot of controversy." He agrees that Davis probably left a year or two too early. "If I had been the manager two years earlier, I'm not sure events would have been any different. I think the denim market and a lot of other things were beginning to click, and they turned around maybe a year after Dan left. Part of it was timing. Maybe part of it was management. Perhaps my style of management was at the right place at the right time," he says. "Whatever the reason, I think we now have the smoothest running companies that we ever have had in terms of employee morale, employee efficiency, and membership satisfaction. If you have all that, you're doing something right. I know our entire staff worked hard the past 12 to 14 years to smooth out our operations, to better serve our members. I think we're seeing the results of those efforts now."

The contrasts between Davis's creative and constantly moving style and Boggs's conservative, bottom-line style seem to extend even to their personalities. At least superficially, Davis appears warm and open, Boggs cool and impenetrable. Both are highly intelligent, Davis in a broad, creative way, Boggs in a deep, analytical way.

Like Davis, Boggs has his admirers and detractors. It's as T. W. Stockton says. "In a cooperative, your friends come and go but your enemies accumulate." To the staff of the cooperative, Boggs is a "hard-nosed" manager, not a tyrant but a demanding, no-nonsense chief executive officer who runs a "tight ship." They sometimes chafe under his style, but they respect him and seem, for the most part, happy and proud to be part of a professional organization that, owing much to his style, functions like a well-tuned machine. The boards to whom he has reported for over 20 years trust his judgment and skill. They have seen firsthand the results of his penchant for precision and his management of the staff and the cooperative's day-to-day interests. The members for whom he works widely recognize his talent and are pleased with the success of the cooperatives under his leadership. Few would ever criticize his management ability; even his detractors praise it, however reluctantly. Some simply do not like him. But where there is dislike, it seems more a matter of style than substance.

Boggs defies stereotypes. He is a West Texan. He grew up on a cotton farm and never completed college. He was an automobile mechanic and Lubbock policeman before studying accounting, first through correspondence courses, then at Texas Tech. But somehow, despite his roots and background, he seems very different from many of the people around him. He is unquestionably a professional manager, comfortable wielding power and dealing with others who do, and always seeming to be in control. He is precise, intense, and ambitious for himself and his organization. He is what some people call "tightly wrapped," not one to reveal his private side and definitely not "one of the boys." His detractors look at his painstaking planning and lack of spontaneity and see scheming. His guarded personality is construed as secretiveness. They suspect that he is *too* ambitious, perhaps identifying himself too closely with the financial and operational success of the cooperatives he has run. Over the years, he carefully guided ACG Board decisions by meticulous research and preparation and convincing presentation. His most severe critics say he contrived many of the board's decisions, "smooth talking" the directors into courses they normally would not have taken.

As long as Boggs's personality is as it is, he probably will remain an object of suspicion and doubt from some quarters. He recognizes "it is impossible to please everyone." When some people say that, it is with an air of nonchalance, as though they just don't care. Not so with Boggs. He is self-assured, but he is not smug. In his more introspective moments, he will admit he cares what people think and, like most of the rest of us, he would like to be liked. But he hopes it is by the cooperatives' record of success, not his personality, that he will be judged as a manager.

Whatever the relative merits of their personalities and management styles, Davis and Boggs seem to have been perfectly suited for their times. When the cooperatives needed ideas and encouragement to embark on new ventures, they had Dan Davis. When they needed conservative management for the new ventures to reach their potential, they had C. L. Boggs. If either had been in the other's place, the results might have been disappointing or even disastrous.

Just a few days before Boggs's appointment, on July 1, Dan Martin returned to PCCA as a field representative after three years, as manager of the Farmers Cooperative Gin at Tahoka. His return would prove helpful to Boggs in later years, as the building and maintenance of good relations with the gins became critically important.

Boggs didn't need the extra work that came with his new PCCA role. In his 10-month-old position as ACG general manager, he faced a mountain of unfinished business. One of the first items that needed to be settled was the matter of the pool. The ACG Board had approved "Plan E" in its July 6 meeting. Now, having made some slight changes, it asked the staff to mail copies to all the gin managers, PCCA directors, and members of the ACG textile pool committee.

In August, Bill Blackledge announced his desire to resign as ACG and PCCA Secretary in order to return to private law practice after 14 years on the staff of the marketing association. His resignation was accepted by the board, "with regret," and Vern Highley was elected to succeed him. It was no coincidence that Blackledge tendered his resignation soon after Dan Davis left. "I'd say that I left because I knew Dan was leaving and I didn't want to stay," he recalls. "C. L. is, I think, a better judge of people than Dan but I didn't really like C. L. that well; I didn't want to work for him." After brief reflection, he adds, "I probably made a mistake in not doing that. I should have, from my own personal standpoint. I would have been better off if I had stayed and worked with him. I can work with C. L. But C. L. is a very ambitious guy, and I had some questions about his method of running the business."

Boggs and the staff were trying both to reduce inventory at the mill and to assure an ample supply of cotton for future operations. He reported to the board that the total 1977 cotton crop set aside for the mill was a distressingly low 29,000 bales. The uncertainty of the number of acres that would remain in the 1978 crop pool and the uncertainty of crop prospects caused him to recommend that those bales be held for the textile mill. It was one of the ironies of that period that ACG was searching, in a worldwide denim glut, for new outlets for its cotton that already had been turned into fabric. And at the same time, it suffered anxiety over whether it would have members and cotton to meet demand when the glut would end, as it inevitably would. The relaxation of pool requirements would not be effective until the 1978–1979 crop, which would not be harvested for several more months. For now, the pool was much too small.

Riding the roller coaster of supply and demand, Boggs and his staff saw their efforts begin to pay off. In June, July, and August,

1978, the denim inventory was reduced by 3.5 million yards, nearly half of that in August alone. Then Boggs reported on a plan ACG might adopt if it became necessary to purchase cotton for operation of the mill. The plan was to contract with PCCA to supply mill needs. Critical to this would be a mutually acceptable transfer price of the cotton involved. PCCA would purchase cotton over TELCOT for the account of ACG. ACG would pay PCCA a "pre-determined fair profit" on all bales selected for mill use. The establishment of the transfer price, and with it the profit to PCCA, was a delicate and critical issue. With the same staff managing both cooperatives, even the suggestion of favoritism toward either side could do damage. Boggs recommended a profit to PCCA of $10 per bale, payment to be made when the cotton was invoiced to the mill. The cotton would be instrument-tested as quickly as possible and that which failed to meet mill standards would be resold. A motion approving the plan was passed early in September, weeks before the board-designated sign-out period would close at the end of the month. The measure was subject to approval by the PCCA Board, and that soon was obtained. The step would assure the mill a larger volume of cotton from which to draw. It was good that it was adopted.

With the end of the sign-out period, estimates of the size of the 1978 pool could be derived. The estimates were not good. The total number of bales that would be ginned at pool gins numbered about 160,000, nearly 100,000 less than the pool gins were supposed to provide. Boggs reported that a survey of gin managers showed about 68,000 bales, or 40 percent of the ginnings in pool gins, would be available to the pool. The survey results later proved to be too low; there would be 85,000 bales available to the pool. But that was still far short of 250,000 bales from which to choose the best 65,000. Instead of selecting one of every four bales, the mill would have to select three of four. Boggs said the size of the pool was "disappointingly low." "Pathetic" may have been a better word. Fortunately, the staff's plan for TELCOT-purchased cotton, approved by the ACG and PCCA boards, would assure the mill would continue to operate without interruption.

As the pool dwindled, the mill flourished. Off-quality production was down to nine percent per week. Denim inventory on September 30 was 4,436,000 yards, some 330,000 yards below projections. And Levi Strauss & Co. wrote that it wanted "every yard the mill can produce for the fourth quarter, in addition to whatever is available from the inventory." It was with pride Bob

Hale reported to the board that, "we are now far and away the largest supplier of heavyweight denim to Levi."

By November, 1978, the denim glut was fast receding into history. Sales of over nine million dollars in October alone led to a drop of two million yards in inventory. LS&CO was clamoring for all the ACG denim it could get. Commitments to LS&CO through March 31, 1979, would leave inventory at a level of only 708,000 yards, down from nine million yards in the middle of 1978. Such are the peaks and valleys of the denim industry. Now ACG faced capacity shortages. It was turning down sales to customers other than LS&CO. It also was trying to maintain the sales network it had built when desperately trying to move off-quality denim, just in case it ever again was needed.

While working to increase production of denim, Bob Hale had the plant in the midst of a stringent cost-reduction program. He told the December board meeting that the cost of manufacturing denim was now down to $1.93 per yard, from $2.00 per yard in the previous year. The program included less yarn devoted to each selvage (specially woven edges that prevent fabric from raveling), a 10 percent reduction in indigo-dye usage, waste reduction in all components, less warehousing needs through reduced inventory, and a concomitant reduction in supply and secondary raw materials inventories. Further quality improvements in the end product, Hale said, prevented returns from LS&CO. Between July 1 and December 1, 1978, LS&CO rejections amounted to only 0.6 percent on shipments of over 11 million yards. Due to process improvements, the number of plant employees would be less than 600 by January 1, 1979. The reduction would be accomplished by attrition, not termination. Hale's objective was to reduce operating expenses a minimum of one million dollars for the fiscal year from October 1, 1978, to September 30, 1979. He was well on his way.

With growing evidence of progress at the mill, the bank slightly loosened its hold. When asked by the ACG staff whether payment of $591,000 in building bonds could be made to gins and regionals, the bank responded that it would "consider a smaller payment." The board also voted to ask for $198,000 for payment of building bonds held by the gins.

The board recognized the contribution made to the plant's growing success by the Crosbyton Gin Division, whose super-gin was providing the plant with the highest quality of cotton it received. The cotton's superior performance was credited to the gin's blender. Tests showed that if the mill were to operate entirely on cotton similar to that delivered by Crosbyton, mill performance would improve to the point that $270,000 annually could be saved, or $3.87 per bale processed. The board put money behind its praise. It voted to credit CGD with $70,817, equivalent to $3.87 per bale on 18,299 Crosbyton bales processed in the mill for the 1977–1978 season. The savings realized by ACG from the use of the Crosbyton cotton were turned right back to the gin. C. L. Boggs was aware that payment of a premium to Crosbyton could create a demand for premiums by other areas engaged in practices such as modular blending. But, he said, in the case of Crosbyton, there is "definite supporting data to the superior performance of its cotton and this should be recognized." The board hoped the payment would be incentive for other gins to improve the quality of the cotton they delivered to the mill.

The following spring, certain a "premium-for-strength" plan would eventually pay dividends in both denim mill operations and raw cotton merchandising, the board directed the staff to develop the details of a program and a schedule of premiums. The staff analysis showed it could be more costly than anticipated, and the board tabled the measure.

Pool membership resurfaced on the board's agenda at the end of 1978. Plan E, adopted in June, stipulated the board would decide by February, 1979, whether future crops would be handled under the 100 percent concept or under an individual producer base concept. The board decided February would be too early, either to make a final decision on whether to return to the 100 percent concept or to offer an alternative program. At a later date, the financial situation of the association might be better demonstrated and producers have more precise information on which to base their decision on signing out or signing in. The matter was tabled.

The staff gathered figures to aid in projection of mill earnings. The numbers would be critical in determining the amount of the progress payment the board would request for 1978 crop deliveries. The board members were told in their last meeting of 1978 that projections seemed to justify a progress payment of seven cents per pound in January. Confident in the staff's work, the board voted to go to the bank with a request to make the payment.

At its first meeting of 1979, the board voted to change permanently the sign-out period from February to April, with April an *additional* sign-out month for 1979. The April sign-out was more compatible with seasonal events that affect decision-making by the producer; it falls after one cotton season and before another. It also would provide more time after the harvest season for ACG to make progress payments that it hoped would convince the farmer to decide to remain in the pool for the coming season. The board agreed that members of non-pool cooperative gins would be allowed to participate in the ACG program, provided they purchased building bonds in the amount of $10 per bale based on the five-year average of a farm's cotton production. In those cases where the production history did not extend to five years, the ACG Board would determine the membership fee.

Some of the previous year's dropouts were trying to get back into the pool for the 1978–1979 crop. Requests had been received from several former members who now wished to re-enter, even though a part or all of their crop already had been ginned. Boggs told the board they were being turned down because to accept them would be to violate the standard agreement and unfair to those who stayed. Eventually, all would have a chance to rejoin.

Fueled by signs that ACG might succeed, rumors abounded that the board would renege on its decision regarding the pool. Word had reached management that "certain producers" feared the board might make a unilateral decision to return to the 100 percent concept. To stem the tide of speculation, the board adopted a resolution that such a decision never would be made without a "grass roots consensus of the pool gins."

With the association's confidence in its denim operations building, the board openly began to discuss what one year earlier would have been unthinkable — expansion of the mill. At the March 8, 1979, meeting, Boggs and Bob Hale raised the possibility of installing 36 additional looms in the weave area of the plant. Playing a card that originally had been dealt by Dan Davis, they noted that space to accommodate the increased number of looms had been provided in the design. Additional looms, they said, would increase production by about 10 percent annually and provide an additional 40,000 yards of weekly production. Six to eight months would be required to acquire and install them.

The board reacted with a question. "Would financing be available from the bank?" The answer was uncertain. Nevertheless, Boggs and Hale were instructed to place an order, to begin the process immediately so purchase and installation could be accomplished as quickly as possible. They were cautioned, however, to seek a 45-day cancellation clause in case the bank would not approve. Within a few weeks, the association had its answer.

The bank was disinclined to finance the 36 additional looms "at this time," preferring to wait, it said, until the association had increased its equity. Same stuff, different day. The directors continued to press the issue. They hoped to convince the bank that the increased earnings from the new looms quickly would repay the cost of purchasing and installing them.

While avoiding the loom issue, the bank finally responded to the request of the previous December for a progress payment to ACG members. The cash payouts the HBC was willing to approve were much lower than ACG would have liked. They included a final progress payment on the 1977 crop of slightly over one cent per pound; a second progress payment on the 1978 crop of two cents per pound; and retirement of one year's building bonds held by the gins in the amount of less than $600,000. It wasn't much, but it was something. And it had been a long time since the members of ACG had seen any return at all. Once the bank decided to approve the payments, it acted quickly. Within two weeks, the board was commending the HBC on its fast approval of the $3.8 million in cash payments to members. Ever so slightly, the pressure was beginning to ease.

The exploding demand for denim had inventory at the plant nearly down to what Bob Hale called "working level," between 500,000 and 600,000 yards. LS&CO's price had gone up to $2.7073 per yard. Labor turnover was down to a record low of 45 percent per year, and the average length of service had increased to one year and five months per worker.

The 1979 annual meeting, held on May 3, was far different in tenor from the one a year earlier. L. C. Unfred called it "one of the happiest occasions in the history of our association. We have moved through a difficult period," he said, "but we have regained strength, and our financial health is good." Many former members were returning to the program in the 1979 crop. Because of these returning producers, Boggs estimated the 1979 crop pool at 185,000 to 190,000 bales, depending on weather and crop conditions. This still was short of the 250,000-bale requirement, but it was far better than before. He also reported on the reversal of denim prices from their extreme lows during early 1978 to the current highs, which allowed prompt approval of progress payments for the 1977 and 1978 crops. The market outlook for denim was "very good," he said, with the trend rising steadily.

Perhaps it takes time for members in an organization like ACG to react to the decisions their elected officials make and the positions they take. At the 1978 annual meeting, with the association in a precarious state, only one director had been opposed. None of the incumbents lost their seats. This year, with good news beginning to be heard from all quarters, a director was voted out of office for the first time.

The board's practice was to meet just prior to the ACG annual membership meeting and then reconvene immediately after. In the early session that morning, Harold Barrett told the other directors there was a possibility he would not be reelected. He was right. He would pay now for his staunch support of the 100 percent concept a year earlier. As was the normal procedure, names had been submitted by the district caucuses, and nominations were taken from the floor. The election was held by secret ballot and the tabulation showed that all current ACG Board members were reelected with the exception of Harold Barrett of District IV. Wayne Huffaker, representing Tahoka Farmers Cooperative Gin, was elected to succeed him.

When the board reconvened after the membership meeting, the directors were Unfred, Stockton, Middlebrook, Whorton, Leach, Brewer, and Huffaker. With Barrett having attended the early meeting and Huffaker the late one, District IV was represented by two different directors on the same day. Among several amendments to the by-laws passed by the membership this day was one providing for staggered terms of three years, with about one-third of the directors to be elected at the annual meeting each year. If the directors' three-year terms were to be staggered, as the membership meeting had voted, then two of the directors would have only one year to serve before running for a full three-year term; two others would have two years to serve before running again; and the remaining three would serve full terms starting that day. Following the membership meeting, the directors decided unanimously to draw lots to determine the terms of office for each director elected that day. The draw showed Leach and Middlebrook having an initial term of one year, Stockton and Unfred having two years, and the rest, including new director Huffaker, having a full, three-year term. The serving officers were reelected unanimously, Unfred as president, Stockton as vice-president, Boggs as treasurer and general manager, and Highley as secretary. A highlight of the meeting and another reversal of recent experience was the board's consideration of applications for pool membership from three ginning communities. It approved membership for Lockney Coop Gin, Farmers Coop Gin Company of Snyder, and Farmers Coop Gin and Supply of Spur. The hemorrhage was stopped; the pool was receiving a transfusion of new membership.

The denim mill also continued to gain in health. As a result, it attracted some unwanted attention. In a telephone meeting on May 14, Boggs told the board that union organizers were contacting mill employees with a promise that they could effect a shorter time period for full vesting for retirement. The program then in effect at the mill provided 100 percent vesting on an employee's tenth anniversary. If the employee left prior to his or her tenth anniversary, the vesting would be lost. The union said it could achieve at least partial vesting on the third anniversary. The staff immediately calculated the cost of ACG matching the union offer. Their analysis showed the annual expense would be between $20,000 and $35,000. Boggs recommended the change be made. It would vest employees at 30 percent after three years of credited service, with an additional 10 percent vesting for each year thereafter, culminating in 100 percent vesting at the end of 10 years. The board unanimously agreed to authorize the change and effectively brought the unionization effort to a halt.

Bob Hale thinks back to the episode, recalling it actually started much earlier and with another company in Lubbock. "Our mill wasn't even up yet. [Unionization] was going on out at Clark Equipment Company. At union meetings over there, it was discussed that [the ACG denim mill] was being built, and certainly they wanted it to be unionized. . . . Then, for some reason, after Clark went out of business, [the union] decided it would be better to wait until our plant was in operation and most of the work force was already here instead of starting right away." Hale says the organizers missed whatever opportunity they might have had. "By that time, we had our policies and our jobs set up, and we were working well with a good group of people that we had trained. By the time they decided to come over here and unionize us, we were already organized. They came anyway. They went to our employees' houses and tried to get them to sign cards. They passed out literature at the gates, the usual things that they do. We had meetings with our employees to explain to them what the situation was. It was the usual tactics for a union and the usual rebuttal from us."

At the June board meeting, Hale reported in detail about the union activities at the mill and the management response. He told the directors he and his staff had held more than 300 meetings with employees in order to deal with the organizing campaign.

Hale talks today about his workforce, some 500 employees. Most of the early employees, he says, came from jobs somewhere in agriculture, "because that's the basic industry here. We had a fairly large number that had been oil field workers, but the majority were tied into the agricultural industry. Some of them had small businesses. "We are about two-thirds male and one-third female, and a lot of wives came to work. Many of our people who had small businesses or farms were able to come to work here and continue their other operations, because here they are working three or four days per week and they still have time to run a small herd of cattle; they can still find time to do their farming operations. They are a general cross-section of the people in the area. . . . Hale says "We found them to be people who adapted very well to textiles, although textiles had never been in the area at all. We are over 50 percent Hispanic right now . . . and the Hispanic-Americans adapt wonderfully to the textile operation, particularly well. They are people who take to training well, and they have a kind of artistic ability that you might overlook at times. They are very dexterous. They have a good work pace and an independence at seeing to their own job and doing it well. All these are tremendous characteristics for a workforce."

Focus on Growth

August, in most places, is a slow and lazy month, less a time for action than relaxation. August, 1979 was anything but that at ACG. The board returned to the subject of expansion, but not simply the addition of 36 looms. The directors discussed the "possibility of expanding textile ventures by ACG." Though they took no action, there seemed to be a consensus that the association should look toward movement into other textile projects. This should be done, the board said, "as soon as is practical, commensurate with building association financial strength at a desired rate, retiring the debt service, and maintaining enough cash flow to keep members satisfied with the program." A year after being on its deathbed, the patient was looking for new races to run.

The staff was directed to study and report on expansion possibilities. By October, Boggs would air a proposal from Kurt Salmon Associates, Inc. (KSA), a national consulting firm, offering a commitment-free, preliminary presentation to the board at its November meeting. Hale lent his endorsement, saying he had met with KSA's representatives and believed them the "best in the business" for this type of study. Later, he and Boggs told the board they felt comfortable recommending that ACG go ahead with Phase I, which would be the identification of products that can be manufactured from West Texas cotton and that have promising markets. The board gave its approval.

The board also returned in August to the subject, which it had shelved in March, of expanding the existing plant. When the HBC was approached then regarding the possibility of financing the 36 looms and said it was disinclined, it also said the cooperative would have to develop a comprehensive plan for modernization before the bank would consider a proposal for additional looms. The staff charged ahead. In fact, it developed two plans within one five-year program. One was a proposal for the addition of the 36 looms to fill the extra space in the original design, the other a proposal for the same 36 looms *plus* a yarn mill and 35 double-width weaving machines. The yarn mill would be built next to the denim mill and provide it with an additional 200,000 pounds of yarn per week. Bob Hale presented the board an overview of the proposals.

Talk at the August board meeting centered on Proposal 1, installation of 36 Picanol shuttleless looms in the weave room. The looms would use purchased yarn, rather than yarn made by ACG. They would cost about $1.15 million to purchase and install, and would have a payback estimated at 0.87 years. The existing weave room layout of machinery and supporting equipment, Hale said, would support the additional looms without modifications to the building. Expected production on the additional looms would be over two million yards per year. Hale said that, with the current market for denim, the extra yardage easily could be marketed. The board voted to resubmit its request to the bank for financing in the hope the request would fare better now that it was accompanied by a comprehensive plan.

Between August and October, the board continued to address the proposed KSA study and the plant expansion plans, and it broached a subject it had not yet tackled in any significant way, management compensation. At an August 10 joint meeting of the ACG Board and PCCA Executive Committee, the main topics were management performance and a profit-sharing plan that had been proposed for key employees of the association.

TEAMWORK

L. C. Unfred opened the discussion with comments about the difficulty of retaining key employees and the importance of keeping them with the association. He gave the floor to C. L. Boggs to review a proposed profit-sharing plan for the eight members of the management team. They were Boggs and Bob Hale; Bert Kyle, then sales manager; Darryl Lindsey, then TELCOT manager; Van Allan May, then controller; Vern Highley, then secretary; H. Mac Cooper, then corporate assistant secretary; and Bill Godlove, director of data processing. Boggs talked about the workload carried by his team. The association employed 26 fewer people in the Lubbock office on July 1, 1979, than one year earlier, and the annual payroll was $288,000 less. Many expenses had been cut, he noted, without adversely affecting operations. He pointed out that several key employees who had left the association had not been replaced, but instead the workload had been spread more evenly over the remaining members of the staff, especially the eight members of the management team.

The concept of the management team was critical to Boggs. "When I decided to accept the challenge of managing ACG and PCCA," he says, "I knew I must surround myself with good people and give them all the responsibility they could handle and the authority they would need to do the job. If I had to pick one management strength I may have, it would be my ability to select and keep a good staff. They are responsible for our day-to-day operations." He described the team concept and said it was working. He said it showed in better employee morale, because the staff felt more a part of the cooperative management, and in efficiencies through closer communication. In addition, the team approach was good for the cooperative because such an approach would make them less vulnerable to the loss of any key employee.

Boggs said the salary level should be competitive with what was being paid in other industries in order to keep key employees. He cited the resignation in May of John Taylor, the association's controller, who left to accept a higher paying job in Dallas. He emphasized that, in order to improve operations and continue to offer the best services to members, "we must keep our key employees." Then he presented a plan for doing that.

A profit-sharing arrangement, he proposed, was preferable to a straight salary adjustment for many reasons. Among them: it would provide an incentive for better management by each person because he shares in the results; the management team members would feel a more integral part of the cooperative because they would share in the good times or suffer in the bad times along with the members; plan participants would make more money only when members made more money and could better afford it. The plan probably would be more acceptable than higher salaries because many of the local coop gins had their managers on a similar commission plan. It could make money for members through better efforts of employees to reduce costs and increase income. And it would tend to reduce employee turnover because employees would forfeit a large amount of their profit-sharing plan if they quit and went to work elsewhere. Wisely, Boggs recommended that the plan be based on cash paid to members and not on net profits, ensuring that employees would have to wait for that part of their compensation until members got their money from book credit (stock) retirements.

Unfred called an executive session and asked Boggs to leave so the board could discuss the issue privately. He said it seemed to him that Boggs's salary "may be low in relation to other managers of marketing cooperatives," and this was particularly true in light of his management of two large enterprises. Comparable salaries at other businesses were discussed, and ACG's independent auditor was asked about options for compensation, including free housing, interest-free loans, and other plans for allowing employers to compensate key employees at a higher level of "after-tax benefits" to them and at less "out-of-pocket" cost to the cooperative.

The ACG Board decided to recommend favorably the profit-sharing plan to the PCCA Board, because both cooperatives would be affected. At an October 10 meeting, Wayne Huffaker moved and Dallas Brewer seconded that the concept of the profit-sharing plan be approved with minor modifications. The motion passed.

A week later, at a joint ACG Board and PCCA Executive Committee meeting, Unfred asked Boggs to present the group information the staff had gathered on the subject. Boggs reviewed the compensation of six of his management team members, comparing them with those who held similar positions on the staff of Calcot, the large California Cotton Cooperative. He left Bob Hale and Darryl Lindsey out of the comparison because Calcot does not have a textile mill manager or a TELCOT manager. The figures showed Calcot's average salaries for their six people were 33 percent higher than the six comparable members of the ACG management team. At a meeting the next evening, the ACG Board reiterated that it approved of the profit-sharing concept, but it "should be set in legal language and further reviewed by the board for final ratification."

The ACG Board and PCCA Executive Committee had approved the plan, but it was for naught. It was defeated by one vote in the 160-member PCCA Board. Though they were not pleased by the vote then, members of the management team feel today it probably was better that the measure was defeated by one vote than had it passed by one vote. Approved by such a narrow margin, it would have been too controversial — more, they feel, than the rewards of the plan would have been worth.

Darryl Lindsey, then manager of the unique TELCOT system, had worked for IBM for five and a half years before joining PCCA in May, 1979. As an IBM executive, he had handled PCCA's account. "Darryl was our account representative when we began TELCOT in 1975," Boggs says, "and he worked to make this system grow from the beginning." Three years after TELCOT was installed, Lindsey was promoted by IBM and moved to Dallas as the company's regional marketing representative in Texas and Oklahoma. "When I became PCCA's manager," Boggs adds, "TELCOT had about a 90 percent reliability rate and a 'downtime' of about 10 percent. In some walks of life, a 90 percent average is pretty good, but not for an on-line computer system. Although Darryl and his family had moved from Lubbock to Dallas only nine months earlier, I persuaded him to return to Lubbock and work for us." Lindsey now oversees all of PCCA's operations, but the one Boggs selects to illustrate his contribution best is the most critical part of PCCA's work, the far-reaching, sophisticated computer network that is TELCOT. "TELCOT," Boggs says, "now has a 99.6 percent availability, which is just one indication of the job Darryl has done."

Its worst days seeming to be over, ACG was receiving many inquiries from producers still wanting to sign additional acreage into the 1979 crop pool at their local gins. Boggs told the board it might be time to review the policy affecting sign-ins, and the advantages and disadvantages of reopening the pool. The directors were not disposed to change. Wayne Huffaker moved to reaffirm policy covering the April, 1979, sign-in period, that any additional acreage signed into the pool following April 30 would be subject to board approval. The motion also said the pool would not be reopened for additional acreage "at this time." It was seconded by Whorton and carried with only one dissenting vote.

EASING INTO MODERNIZATION

Sometimes, the ACG Board, like any other committee, was a clumsy creature that couldn't get out of its own way. Some issues seemed to drag on forever – constant, repetitious discussion followed by the tabling of the issue for reexamination at a later date. There were times, however, when the board could move at breakneck speed, especially if it seemed to sense an opportunity at the other end. The November, 1979, board meeting is an example. Bank approval had just been received for the purchase and installation of the 36 looms. With that in hand, Boggs reported an order for the looms was being placed right away. Hardly missing a beat, the board began discussion of the positive effect that Expansion Proposal 2, the more comprehensive of the two, would have on contract negotiations with Levi Strauss & Co. Having just received the bank's permission to buy 36 more looms, the board was looking at the possibility of building a yarn mill and adding 35 *more* looms. Boggs commented, almost offhandedly, on the sharp trend toward shuttleless looms in the textile industry, and he called on Bob Hale to elaborate. Hale said comparisons of shuttleless looms and the older, standard fly-shuttle machines showed the former were superior in efficiency and quality performance. ACG, he said, would need to "change out its shuttle looms at some point in order to remain competitive in quality." The change need not "occur overnight," Hale said, "but the board should give thought to gradual replacement."

Two bank representatives attended the meeting. The Houston Bank for Cooperatives had been relocated to Austin, the state capital, by the Farm Credit Administration, and now was called the Texas Bank for Cooperatives (TBC). Though it is not a matter of record, the bankers must have been incredulous at what they were hearing. Their approval of 36 new looms had mushroomed into a discussion of building a yarn mill and more looms and possible replacement of all the looms in the mill.

So the point wouldn't be lost on the bank officers, the staff asked the board if it favored Expansion Proposal Number 2, provided financing could be obtained. The answer was "yes." The bank officials said neither the Texas nor Central Bank "would be inclined to approve Proposal Number 2 of the expansion project until the Association's debt-to-equity ratio goal is reevaluated." David Dabbs said the TBC "is not against the expansion project but approval would be difficult at the present time." ACG's point having been made nonetheless, Boggs drove it deeper. He said he did not want to take on additional projects, but that "thought has to be given to minimizing the risk of facing the future without a new long-term contract with a major denim user of the caliber of Levi."

At the December board meeting, the last of the year, it was announced that a contract had been signed for purchase of the 36 Picanol looms. Arrival of the looms was expected around June 15, 1980, and it was anticipated that they would be in production around the first of August.

On January 9, at the first board meeting of 1980, Boggs reported on discussions he and Bob Hale had had with Levi Strauss & Co. in San Francisco. The talks regarded extension of the five-year sales agreement that was due to expire on December 31, 1981. LS&CO's response was positive. Though it was no longer their policy to consider commitments beyond a three-year period, they would extend the contract. As part of the agreement, however, they wanted to see ACG modernize its facilities in order to improve the quality of its denim. With that condition in mind, the board further intensified its focus on modernization, though it moved very cautiously.

Jack Hughes, president of the TBC, was at the meeting. By this time, the association had no trouble getting the TBC to authorize progress payments, and Hughes's participation in ACG board meetings no longer had a confrontational edge to it. The directors voted to request approval of an immediate progress payment in the amount of $4.9 million. There was little doubt it would be done. Now the board was interested in Hughes's reaction to the modernization issue. ACG had begun to see modernizing the denim plant as being critical. The board felt one of three things would happen if they did not modernize, and all of them were bad. Either the plant would not have the increased capacity and improved quality to secure an extension of the sales agreement with LS&CO; or the association would have to make major concessions to LS&CO; or it would have to find a way to market denim without a LS&CO contract.

Hughes spoke for the bank. He said the TBC was not opposed to modernization, but that ACG needed a concrete proposal and long-term commitment by its members. If money were going to be borrowed for modernization, Hughes clearly wanted to see the patronage of the pool locked into a commitment. He contended, too, that there were major advantages to be enjoyed if the regional cooperatives combined to "group their assets." He talked at length about the economic and operational benefits such integration would afford. It was a subject Hughes would emphasize many times in subsequent years.

The board members were attentive. Their comments recognized the validity of his ideas. But they also knew how controversial and intensely political was the nature of what he proposed. Though the regionals were kith and kin, each prized its independence. In any case, the ACG directors were not there to talk about strategic plans for the regional cooperative system. They were intent on doing whatever was necessary to get an extension of the LS&CO contract under conditions favorable to the cooperative. No decisions came from the January meeting, but in subsequent weeks, discussion of Expansion Proposal 2 faded. Talk increased about the idea that Boggs and Hale had cast out to the board almost casually at the end of the preceding year, that of converting entirely to the new shuttleless looms.

In February, Kurt Salmon Associates completed Phase I of its study and reported to the board at a special meeting. Jim Hicks of KSA said the firm had studied many textile products and areas for compatibility with ACG's operations and West Texas cotton. It focused on three product groups: corduroy, terry cloth, and denim; and its conclusions pointed to denim as having the "strongest opportunity for any contemplated expansion of ACG's textiling program." Denim, in one form or another, would continue to be the cooperative's "bread-and-butter" product. The report, along with favorable comments from the bank on the five-year plan the staff prepared, launched another discussion of modernization options.

Proposal 2 was complicated, involving new looms, weaving machines, and a yarn mill. Bob Hale said more complete data would be needed to project a payback period. Many variables remained to be studied, but he said "reasonable" figures had been put together for the board to decide if it wanted to continue exploring that option. The directors still were undecided. It clearly was a major decision, and concern weighed heavily on them. On one hand, they wanted to go ahead. They were afraid they would not secure an extension of the LS&CO contract unless some type of modernization could be promised. They also had in mind the findings of Phase I of the KSA study showing increasing opportunities for heavyweight denim well into the decade.

But also pressing on them was the very real possibility that they would not be able to sell the project to their members. And they were concerned about whether they could raise sufficient capital to satisfy any request for investment funds by the bank. Several directors said they just did not have enough information on which to base a decision. They needed more details on the full cost of the project and its potential impact on cash flow. Someone suggested extensive studies. Someone else, taking the idea further, suggested they go back to Lockwood Greene for a more refined study and projections. Such a study, the staff reminded them, might cost $25,000 to $50,000. If a decision were made to proceed, an engineering study might cost another $500,000 to develop plans ready for bid-taking.

Their concerns about cost caused them to remember the words of Jack Hughes. Perhaps the regionals ought to be asked to assist in the initial funding. They mulled over the possibilities. Membership would have to be opened to producers in all cooperative gins, so a campaign would be undertaken to expand participation. And with that kind of increase in participation, it might even be possible to get an additional loan guarantee from FmHA. But then, old reservations resurfaced. Several members emphasized "the obligation of the association to the original and current members," and raised the issue of "whether it would be fair to them to expand the association without requiring a comparable investment on the part of new members."

The board could not seem to make a decision on how to proceed on Proposal 2. But it could not seem to table the measure, either. A motion was made to discontinue further consideration of the proposal. It failed for lack of a second. The better answer it was felt, was to study it more. It was moved to extend the study being done by the staff, further analyze the costs and benefits, and estimate the impact of having to market denim in the event an extension of the LS&CO contract could not be secured. This motion failed, two votes to three. A third motion was offered, directing the staff to take the current cost study and analyze it with respect to impact on the cash flow of the association. The staff was told to conduct its study "in a simplified manner." This might be called the "Let's study it, but not too hard" approach. That motion expired for lack of a second. Finally, in the realization it was going nowhere, the board agreed to discontinue further study of Proposal 2. The plant modernization issue would continue to receive attention from the staff, but it would lie dormant in the board until the 1980 annual meeting in June, more than four months later.

David Dabbs of TBC had resigned and was replaced by Don Scott as loan officer for ACG. Scott had been closely involved with ACG since its inception, and so the change would cause no disruption in the relationship between lender and borrower. Because of the turnaround in the denim end of ACG's business, Scott would have a considerably easier time than his predecessor. At least he would not have to say no as often.

Though the pool was not nearly at that level, the board voted in March to limit it to total acreage that would produce 300,000 bales. A month later, Jack Hughes told the directors the TBC and FmHA both would like to see the pool back to a level of 300,000 bales. From all points of view, that seemed to be the ideal size. The board's response was that the pool might approach that level if the 1980 crop was not "hailed out" as was the case with many producers in the 1979 crop. The idea of permitting growth in the pool to raise capital for modernization seemed, at this point, to have withered.

In the days leading up to the 1980 annual membership meeting, the board decided the titles of ACG's officials were not in keeping with their responsibilities or the size and scope of the organization. They proposed the titles be changed to bring them more in line with the rest of the corporate sector and other major cooperatives. The change would require an amendment to the bylaws, and so the membership would have to approve it. Under the plan, L. C. Unfred and T. W. Stockton would be chairman and vice-chairman of the board, respectively, titles much closer to their actual functions than president and vice-president. The general manager would be retitled as president. For Boggs, who in reality was chief executive of the cooperative, it also was an appropriate change.

FORCING THE ISSUE

John Hunt, then manager of the Regular Jeans Division of Levi Strauss & Co., was the keynote speaker at the annual meeting on June 19. He had returned from his tour in Hong Kong and resumed a close relationship with ACG. As in years past, he continued to make himself available to the association, visiting Lubbock when ACG would call him because its leaders felt an LS&CO presence was necessary for support in their dealings with bankers, government officials, and even their own membership. Hunt's presence that particular afternoon was critical. Before the board could decide what course to take on modernization, the members had to be convinced it was necessary. And before they would consider any modernization of the mill, they needed to hear that prospects were good and it would be a sound investment.

Hunt says he wrote some notes on the airplane as he traveled to Lubbock. When he arrived, C. L. Boggs had some points in mind that Hunt should make. He handed Hunt some notes, asking him, "Would you mind saying this?" "C. L.," Hunt replied playfully, "why didn't you just write my speech for me?" Understanding the importance of what the ACG staff was trying to do, Hunt gladly agreed to help. As he says today, "We'll always be committed to ACG. It's one of those relationships we're so proud of. And my job was always to help them if I could."

In his remarks, Hunt praised the relationship between LS&CO and ACG and the negotiations that resulted in their original contract. And he paid tribute to the late Roy B. Davis. Then he reviewed current denim prospects. He noted that the existing sales agreement between LS&CO and ACG had 18 months of its five years remaining. He told the members LS&CO was willing to extend it an additional three years, "if the association will modernize its facilities. We can't help you in the financing, but we can continue to provide you a market," he said. "In any event, we must find the quality of denim that we need." The message to the members and the bankers could not have been more clear. LS&CO would extend the contract if ACG would modernize its facilities to improve quality. Hunt admits, "I was down there to make the speech supporting this."

Immediately after Hunt finished, Bob Hale reported on the installation of the 36 shuttleless looms. The purpose of the new looms, he told the members, was to make ACG more competitive against mills that already had installed the technology. Then he and Boggs stunned their audience. They announced the board was studying the "need to replace all the present looms with the new looms." The conversion, it was said, would cost about $10 million.

John Hunt recalls, "Bob did an excellent selling job. He had to go before all these [ACG] members and say, 'The loom we have today was good since before the time of Christ, but with all the new equipment coming on the marketplace, and in order to [improve our quality], we have to adjust to the industry.'" C. L. Boggs and Bob Hale told the members that the 36 new looms would increase production about 10 percent. This was far from the 30 percent Hunt said LS&CO was looking for. But implied was that many more of the new looms would expand production to the level LS&CO wanted.

The membership did not act directly on the modernization issue, nor was it expected to. Normally, the only issues put to vote in the annual meeting were changes in bylaws or election of directors. What was important was that the membership understand the issue and support the board. And it did. Hunt vividly remembers, ". . . all these people in the audience saying, 'Go ahead! What are you waiting for?' I looked around," he says, "and I just couldn't believe it. I said to myself, 'What the heck . . .?' These farmers were just totally committed."

In the annual election of directors, the membership would vote to fill only two seats, having voted the year before to change to staggered, three-year terms. The results of the District I nominating caucus showed Doyce Middlebrook the only nominee. He was reelected without opposition. Due to "personal demands," Joe Leach had announced he would not seek reelection from District II. Albert Sammann and Levon Harmon were put forward by the District II nominating caucus as candidates for Leach's seat. Jim Bob Curry was nominated from the floor, pitting him against the two. A run-off was necessary between Curry and Sammann. Another secret ballot was taken and Curry was elected.

For certain District II members, the meeting and election would be their first. Several months earlier, the board had voted to assign all ACG participants from non-pool gins to District II. This would give them an affiliation for participating in annual meetings and nominating a district director.

One week after the 1980 annual meeting, the board held its monthly meeting for June. As in every year, L. C. Unfred declared all offices vacant and called on Vern Highley to conduct an election. The nominees, all of whom were elected, were Unfred as chairman, T. W. Stockton as vice-chairman, C. L. Boggs as president, Vern F. Highley as secretary, Van Allan May as treasurer, and H. Mac Cooper as assistant secretary. On staff, Darryl Lindsey was named vice-president for operations; Bert Kyle, vice-president for marketing; and Van May, vice-president for finance. Bob Hale also enjoyed a change in title. Formerly called manager of the textile division, he was promoted to the newly created position of vice-president for textile operations.

With the annual meeting and elections behind it, the board again broached the issue of plant modernization. It was the one issue, more than any other, that would dominate the association's agenda for the rest of the year. Although they agreed in February that modernization did not seem feasible, due to the large capital requirements involved, the directors had heard John Hunt's message. No modernization, no contract extension. They had more questions for LS&CO; they wanted to be absolutely certain of their options. Undertake expansion? Replace the present looms? Not do anything at all? And what was the impact of each? The board had gone on record in favor of some sort of modernization, subject to the availability of financing and membership approval.

But that was mainly aimed at eliciting an opinion from the bank, which to this point only had said it would not commit to anything until ACG made a decision and a formal request for funds. The board was caught between LS&CO, the bank, its members (only a portion of whom were at the annual meeting), and its own indecision. And it was desperate for answers.

Within a few days, the bank reacted to the board's vote in favor of modernization. The conditions it put forth in the letter were hard. In particular, the bank said it would prefer extension of the marketing agreement between ACG and its members to a period of five years. The bank implied some of its conditions might be negotiable, but the board was more discouraged than ever.

Despite misgivings of the bank, the ACG staff continued its research. Bob Hale was obtaining as much information as possible on the various types of shuttleless looms. The two that looked most promising were those manufactured by Picanol, such as the 36 recently purchased, and by a firm called Sulzer. In September, he was back before the board, talking about the relative advantages and disadvantages of each. The best news Hale brought with him was in a letter from Waverly Watkins of LS&CO's quality control center, in which Watkins compared the "old" and "new" styles of denim produced by ACG. The old was produced on the shuttle looms; the new on the shuttleless looms. Watkins said he was "elated" with results from the new production and asked how soon the denim mill might have its entire production from shuttleless looms. Hale said it would take 18 months to two years before delivery could be taken on the Sulzer looms. Picanol looms, however, could be available in about seven months.

By October, the ACG staff was pushing hard for the Picanol looms. C. L. Boggs said the staff felt Picanol offered "numerous major advantages to ACG when compared with other machines." He cited lower costs, faster delivery schedule, ease of maintenance, and design of the ACG plant, which better accommodated the Picanol loom. He also pointed to another advantage — Levi Strauss & Co.'s preference for the quality emanating from the 36 Picanols in the mill, now a matter of record.

The staff continued to research the economics of conversion, as well. They projected costs for purchasing and installing 338 new looms were about $13.3 million. ACG might recover $1 million to $1.8 million from sale of the old fly-shuttle looms. Estimated increase in net profit from the new looms was $3.5 million to $4 million annually based on economic conditions. The payback period would be about three and a half years. Hale and Boggs said they would try to negotiate a deal with Picanol executives at a textile machinery show later that month in Greenville, South Carolina. They asked to be authorized to sign an order for the looms if acceptable terms could be worked out, with the complete understanding that it would be subject to confirmation by the ACG Board and the TBC.

At this point, the board was doing everything it could short of having money from the bank and an extension from LS&CO in hand. It would not take a plunge. But, using the staff to help identify the options and clarify the implications, it was moving faster to action. Several directors were determined that the staff understand any agreement was subject to board approval. In its first four years, ACG had been through some very rough times and only recently was seeing any relief. Like a spouse who had been through a bad marriage, its directors viewed another trip down the aisle with some apprehension. Nonetheless, they voted to give management authority to place a tentative order.

At the October board meeting, Bob Hale reported the tentative order had been placed with Picanol, subject to formalization within 60 days. Picanol agreed to ship the first looms from its Belgian plant not later than March 15, 1981. They agreed to lower the price of the looms by about $400,000. ACG would have until delivery of the new looms to sell its fly-shuttle looms. The number of looms unsold at that point would be purchased by Picanol for $4000 each.

Hale stood rock-solid behind his recommendation, that the Picanol shuttleless offered greater advantages to ACG than other brands or technologies. He predicted that, by the end of 1981, over 90 percent of LS&CO's denim would be manufactured on the new looms. The board had heard enough. It had the expert opinions and forecasts of Bob Hale, C. L. Boggs, and the rest of the management staff on which to base its move. Hale already had proven himself to be far more than the "denim man" the board had hoped he would be when they hired him. He was a first-rate textile engineer and plant manager. Boggs says, "I haven't met all the textile engineers in the world, but I don't think there is one better. There may not be any as good."

Boggs, Hale, and the rest of the management team had established a reputation for exacting research and analysis and dynamic follow-through on board directives. It was time to put aside all the talk of various options and vote on conversion of the mill entirely over to new looms. T. W. Stockton introduced the critical motion.

The board voted its approval to proceed with the project. It would require bank financing and be subject to acceptance by the membership, but the board finally had put on record the direction in which the association would move. The vote took courage. Many at the June membership meeting were in favor of undertaking a modernization program, but they were only a portion of the total membership, and it could be argued they were agreeing with modernization on principle. This board vote was on spending $12 million after years of hard times.

Jim Bob Curry is a well-liked and thoughtful man, sensitive to the desires of the farmers who elected him. He recalls "I had just been on the board a short time when we faced this [prospect] of having to spend over $10 million putting in new looms. I remembered some of the friends that I knew down at Tahoka and Lynn County. After I went on the board, they said, 'We don't want you to spend a big bunch of money. We want you to conserve.'" He describes the mixed feelings he had about the decision. He acknowledges the price tag for the new looms was high, but adds, "We [ACG] were having a terrible time getting along. We had too many seconds, and production wasn't up to good standards. I felt like about half [the people I represented] were for putting them in and the other half not." With his constituents split on the issue, it was a hard decision for Curry. It was no easier for the rest of the directors. But they sucked in their breath, crossed their fingers, and made it. "I believe to this day," he says now, "it was the best move we ever made. We cut our seconds down from 18 percent to one or two percent, an awfully good quality denim for West Texas." He smiles an easy smile of satisfaction. "Made you feel kind of proud that you went ahead and voted for it after it turned out like it did. It turned out real well."

Wayne Huffaker feels the same way. "We got flack from the membership over spending that $12 million," he says. "The board, for the most part, was mixed in its feelings." Describing himself in a way that many of his colleagues would, he adds, "I'm one of those guys that once I decide something is right or wrong, I take the side I think is right and I'm going to fight you until hell freezes over. A lot of those guys," he adds, "are just not turned that way. They sit back and listen . . . and probably do a better job than I do." Because some directors were quiet on the issue does not mean the decision was easier for them. Huffaker says there was no vigorous opposition on the board to the looms, but a lot of reservations. "In the country, in the membership, when the word got out that we were fixing to spend $12 million when we just spent $50 million building [the mill]," he says, "there was a lot of criticism. . . ." The mill cost less than $50 million to build, but Huffaker's point still stands. In sum, he adds, "It turned out to be a great investment."

Robert D. Haas, now chairman and chief executive officer of Levi Strauss & Co., considers the decision to purchase the new looms a landmark in ACG's history and its relationship with LS&CO. Though he was not regularly involved with ACG at the time, he was "aware of the very courageous decision that was taken relatively shortly after [ACG] came into being to completely scrap their huge investment in traditional shuttle weaving machines and go to state-of-the-art shuttleless weaving machines. "Given the infancy and fragility of the enterprise," he says, "it was an astonishing decision. It was typical of the foresight of the West Texas cotton growers who banded together to create ACG that they would be committing to the long term rather than trying to achieve [their] break-even [point] and profitability as quickly as possible. That was a really important signal to all of us within Levi Strauss & Co. that these folks were serious and committed to our success just as we were committed to their success."

Jim McDermott of LS&CO shares Haas's admiration for the ACG Board and its ability to act rapidly on a decision once the options were clear. "When they came up with something, when they came to a conclusion, they really moved. Those guys," he says, "had guts!"

C. L. Boggs told the board it would meet with bank representatives on November 3 to discuss financing the conversion, and he reviewed the projections and assumptions to be used in talking with the bankers. Assuming ACG would have to go to FmHA for another loan guarantee, the staff was covering that base, too. Boggs told the board he had visited with Jack Boyd and Don Perry of the Texas and Lubbock FmHA offices, respectively. He and Bill Blackledge, then acting as outside counsel to ACG, would meet with FmHA national officials in Washington on November 4. Boggs expected ACG's application to be completed before that meeting and approval by FmHA to come within 60 days.

On November 19, Boggs brought good news from his discussions with the bank and FmHA. The bank would not require a supplemental loan guarantee from FmHA for the conversion project, provided FmHA would approve a subordination agreement to the original guaranteed loan. With this subordination, the bank would consider financing the entire project, about $12 million, if ACG would increase its working capital to $10 million by September 30, 1981. Even without the subordination by FmHA, Boggs said, the bank would consider financing 80 percent of the project. If ACG met its projections, it could respond to either circumstance, by increasing its working capital to $10 million or financing 20 percent of the project itself. To protect the association in case the FmHA subordination could not be achieved, the board voted unanimously to apply for a $12 million term loan from the Texas Bank for Cooperatives to cover financing of the project.

At the same meeting, Bob Hale buttressed the decision of the board by reporting a comparison of the new looms and the old. The efficiency of the shuttleless looms was 93.2 percent in the prior week; fly shuttle efficiency averaged around 85 percent for the month to date.

One week later, Boggs called a special meeting to tell the board that he and Hale had met with Levi Strauss & Co. officials and a tentative agreement had been reached on the contract extension, subject to approval by the ACG Board and other LS&CO executives. The contract would be extended three years, through December 31, 1983. With the Levi contract falling into place, the pace of work on the financing gained momentum. On December 10, Boggs reported on actions needed from the board which, he said, had to move "on a fast track." They had six days, to be exact. Picanol had offered an attractive sales price and early delivery of the equipment if a December 16, 1980, deadline was met. It later extended the deadline 60 days on the price, but ACG hoped to bring the issue to a close by December 16 to ensure early delivery. In the meanwhile, FmHA informed ACG it did not have $12 million remaining in its current-quarter budget and, therefore, could not approve a supplemental loan guarantee in time for ACG to meet its contractual deadline.

The ACG Board and staff discussed the working capital position of the association. Assuming all projections for the 1981 fiscal year were met, surplus working capital above $8 million would yield about $2.3 million, leaving $100,000 to be raised in meeting the $2.4 million necessary for ACG's 20 percent part in financing the project without the loan guarantee. The board requested the TBC to commit in writing to 80 percent financing, or that a representative of the bank be present at its next meeting so the commitment could be made official. A motion carried unanimously to authorize the staff to confirm with Picanol the purchase and delivery agreement of equipment and supplies. The staff was further instructed to try to obtain a minimum two-week postponement in the down payment, but, if not possible, to proceed with the down payment. Picanol granted a three-week postponement.

At the December 17 board meeting, Boggs reported the bank had agreed to a collateral-sharing subordination plan to finance the shuttleless loom conversion project, because the FmHA loan guarantee was not feasible "due to budgetary limitations." A telex from the TBC outlined the bank's agreement to provide 83 1/3 percent of the amount of the project above the $1.4 million sales price for the old looms, to a maximum of $10 million. Picanol received its down payment. The deal was done.

In its final actions for 1980, the board took up the cases of two ACG producers, one of which had been under review for several months. In its handling of the two cases, the board demonstrated both discipline and compassion. The first was discussed at a fall meeting from which two directors were absent. Those present considered it such a "serious problem" that action was deferred until all were present. Farmer H., from Friona Farmers Cooperative Gin, said he instructed someone at the gin to void his contract for the current year on his three farms totaling 838 acres. But there was no written record. The gin submitted a letter supporting his contention and requesting that he be released from his contractual obligations for the 1980–1981 year. There was general agreement by the board that the matter could pose a serious challenge to the credibility of policy governing membership sign-outs. The farmer's claim and the support it received from the gin would not sway the directors. The board moved unanimously to enforce action against him, including legal measures if necessary, to abide by the ACG contract, "since there is no written record of his having signed out during the regular sign-out period in April, 1980."

At the same time, the board moved unanimously to release the estate of H. Joe Schwartz, Jr., from the pool. Schwartz was a young farmer in the Hale Center Cooperative Gin who died in a farm accident. The board, although intent on adhering to the letter of its agreement with producers, had no desire to add to the grief of a member's family by dragging them through legal proceedings.

On the last day of 1980, with a winter chill in the air, the board received warming news. The case involving Farmer H. had been peacefully settled. He signed ACG's memorandum of agreement for delivery of his cotton. The problem was resolved. Happy New Year.

Trouble in the Family II
CGD Versus ACG

The major story for ACG in 1981 had nothing to do with Levi Strauss & Co. or the denim mill at Littlefield. Something ominous was happening within the ACG family itself. Like a slowly spreading virus, a strain in relations between the Crosbyton Gin Division and the rest of ACG was growing. It started long before 1981, though no one seems to recall an epochal event that caused the rift. The relationship deteriorated a little at a time, almost imperceptibly; 1981 was when it showed clearly in the deliberations of the board. For a year, little else would move it off the agenda.

The conflict was one of many ironies in ACG's history. The Crosbyton Gin Division, also known as District VII of ACG, was part of American Cotton Growers. But it was also the parent of ACG. Its members were the nucleus of the cooperative; their investment included the pledging of virtually all their assets. And they were not happy with the way they were treated. There was a broad foundation underlying CGD's dissatisfaction, but their outspoken point of contention was displeasure with the accounting methods and procedures that were being used.

Van May says the complaint was justified. Though the problem started before his arrival on the staff in March, 1978, he talks about the genesis of it. "[PCCA] had grown from being just a cotton merchant to TELCOT, and then to more and more gin services, on-line computer services, and so forth. ACG had grown from just a dream that somebody had to a 300,000-bale pool and a textile operation. Anytime you have dramatic growth in a company," he says, "the systems, the record keeping, usually lag behind. When I came here, that's the way it was, 'snakes' under every rock. Every account you looked into had a snake in it, a lot of problems which developed into some real credibility problems for ACG over in Crosbyton, because this office was responsible in those days for keeping the Crosbyton Gin's records. And they were in as much of a state of disarray as everything else." He quickly adds, "It was no one's fault. It had kind of outgrown everybody. The staff here was at one level of expertise for this small, 'everybody-knows-what-everybody-else-is-doing' company that had suddenly grown to multifunction." He says the "bizarre structure, with several entities all tied together," compounded the problem.

May had been on the staff for about one year when his supervisor left. C. L. Boggs says today that he had a "difficult decision to make in deciding whether to promote Van to the position of treasurer or go outside and bring in a more-experienced person. He convinced me I should give him the opportunity. He said if he could not handle the job, he would step aside and I could bring in someone else. How could you not go with someone with that attitude?"

May describes the period from the time he joined the staff to when he assumed responsibility for all the accounting and record-keeping functions. "We had forest fires raging out of control. Whichever one scorched the worst was the one we tried to put out. There was a lot of disarray. We had set about getting all that fixed as fast as we could, but you never get things fixed as fast as people want. The majority of them were resolved within a couple of years. By the early part of the eighties, we had a pretty good handle on what was going on." But by then the damage had been done. "A number of things had happened during that period," May recalls, "which led the Crosbyton group to want to split off and not stay a part of the corporate entity of American Cotton Growers. The Crosbyton group, [which had] pledged all its assets and become part of [ACG] as expanded, did not feel they had been given adequate or reliable financial data to manage their part of the operation."

In the February, 1981, board meeting, T. W. Stockton said the Crosbyton Gin Division's board of directors was not in agreement with financial reports prepared by ACG's accounting firm, Edwin E. Merriam and Company, for the fiscal year ending September 30, 1979. It was agreed the two boards should discuss the matter jointly at a later date, and on February 24 they convened for that purpose. C. L. Boggs prefaced the meeting with a review of background information leading to the organization of ACG and construction of the Crosbyton Gin in 1973. Stockton, as vice-chairman of ACG and chairman of the Crosbyton Gin Division, took the floor. Stockton had a foot in each cooperative. This time, he was speaking for his home operation. He talked about development of the Crosbyton Gin, including difficult and costly experiences they had had in getting certain equipment, such as the blender, to perform to acceptable standards. They had faced enough problems just getting the super-gin running, then they were called upon to take responsibility for the textile mill.

By early in 1981, CGD had developed a bill of particulars regarding sacrifices and hardships it had endured in the name of ACG. Stockton was joined by other CGD directors — Compton Cornelius, Clyde Crausbay, J. K. Edinburg, Noble Hunsucker, D. J. Moses, and Charlie Wheeler — in bringing issues to the floor. They talked about the progression of financial outlays, distribution of dividends to patrons, pledging of the gin's equity as collateral to borrow funds to build the denim mill, effects of changing the fiscal year, and other circumstances of the past seven years.

Van May remembers that Stockton was "really on the hot seat" during the entire episode. "He was in the position of making statements that he was instructed to make by his board in Crosbyton that were pretty inflammatory in the setting of the American Cotton Growers' board room. He didn't relish making them." Asked to speculate on Stockton's sentiments, May says, "He spent a few hours a month in [ACG] and the rest of his life at Crosbyton. You have to live there, with people you have known and worked with all your life." He quickly adds, "I'm not saying that some of the issues they raised were without merit, either. I think some of them had real merit, and so it was pretty easy that he should feel strongly and lean toward the Crosbyton side of the issues. On the other hand, he did a really good job at balancing that with his responsibilities as a member of [ACG]." As if to make sure the point is not lost, May adds, "He really did . . . just a tremendous job."

Wayne Huffaker, who came to be the leader of the ACG faction, remembers Stockton's role from his vantage point on the other side of the issue. "He was in to represent Crosbyton in the way their board and their manager were telling him to represent them. I never did hold any hard feelings against T. W. In fact, I consider him a good friend." Asked about Stockton's conflicting loyalities, Huffaker quickly answers. "I felt like he played fair on everything. His views just differed from mine and some of the others on the board."

Today, Stockton says he never heard too much complaining from the Crosbyton Gin about them having all their assets pledged to the mill, "until we started having the [American] Agriculture

Movement. We had some [Crosbyton] members who were young and involved in it, and they used about anything they could to find fault.'' Some of these young producers ''were uninformed about what was really going on. They were working with information that was not accurate. But they had been told, 'This is the way it is''' by some of the movement leadership. ''It got down to the point where you either just let it ride or you challenged the information they had. I chose to challenge it. And we had quite a battle in [Crosbyton]. We lost some of them,'' he adds, ''but they came back.''

L. C. Unfred, acting as the peacemaker, said the ACG Board would be willing to explore ways in which the problem could be resolved. Both boards agreed on the need to have accuracy in the financial reports and to take whatever steps might be necessary so that everyone would be satisfied with their accuracy. They imposed a deadline, resolving to reach a ''workable solution'' prior to the Crosbyton Gin's annual meeting.

Then, according to the minutes, there was a ''brief discussion addressing the question of whether the Crosbyton Gin Division eventually should separate from the central association and operate independently as a cooperative ginning association.'' It was open talk of a split in the ACG Board of Directors.

It is important to understand that this was very much like a family quarrel. There was a serious conflict and talk of a ''split'' or a ''spin-off.'' It was even referred to as a ''divorce'' at times. But it never would have resulted in a total separation. It meant, instead, a separation of the two entities as corporations with the same accounting system. If it was a ''divorce,'' it was one where the parties would live in the same house and sleep in the same bed, but where the property would be carefully divided and individually owned and accounted for. Though anything is possible when emotions run hot, it is hard to imagine that anyone ever seriously entertained the idea that the two cooperatives would sever all their dealings. They had too much in common and too much at stake for that to happen. The Crosbyton Gin Division would continue to be a part of American Cotton Growers, albeit under another name and with its own set of books. Nonetheless, as in family quarrels, emotional ties made matters worse.

Clyde Crausbay, a Crosbyton Gin Division board member, recalls it felt to him like CGD was a ''child in the family. It seemed that [the relationship] should have been smooth. We never asked for anything that they didn't give us. But maybe,'' he adds, "we're a bunch of individuals and just felt like we had to ask the 'big board' for everything. "Just a year earlier, for example, in February, 1980, the Crosbyton board had to get the ACG Board's approval to purchase 25 module builders and support equipment for $667,000. The bank demanded strict accountability from all the ACG participants. But that was small consolation to the CGD members. Intellectually, they understood it. Emotionally, it was hard to accept. It must have been galling for them to have to go, hat in hand, to the group they had created in order to seek approval.

Van May describes it from the Crosbyton point of view. "Everything they did out there had to be cleared with the 'big board' here in Lubbock, the seven-man American Cotton Growers Board, and there are a number of things that . . . involve hard feelings on a deal like that. . . . There was the typical local group versus corporate group friction, even though American Cotton Growers never once turned down a recommendation from the Crosbyton committee. They never turned them down, and yet it still rubbed the Crosbyton group the wrong way that they even had to ask." From the ACG viewpoint, he adds, "But with . . . this operation, which was really floundering in 1977, 1978, only barely beginning to pull out of it in 1979, obviously everything had to be very closely scrutinized and the bank had to approve everything we did all the way down the line. [Crosbyton] never liked that, and it's not hard to understand, it's just human nature. You prefer not to have it that way."

He cites another problem. "The multiple gin division concept never came to fruition. [Crosbyton was] the only gin who had their assets pledged; everybody else had $10 a bale pledged. If ACG went down, they lost their gin and everything. If ACG went down, these other gins lost some part of $10 a bale. Big difference. . . . And they didn't feel they had ever been compensated for that difference," May says. "The people who bought building bonds were paid interest; Crosbyton had never really been paid interest on their equity that they had put into this operation. . . . So they didn't like the setup. There obviously were not going to be other gin divisions like them; they had everything pledged, more than any other participants had." That feeling caused bitterness on the part of some Crosbyton members. In addition, May says, "they felt like they were getting unreliable financial information from this office to run their business."

"I think American Cotton Growers was having a lot of trouble with bookkeeping, and it seems like we couldn't tell where we were," Crausbay recalls. "We just didn't feel comfortable when we got a statement, whether it was right or not." When asked if Crosbyton's higher exposure to risk was a factor, he adds, "Yes, we felt that we'd put up everything. We had it all to lose. And the other gins . . . a $10 [per bale] bill . . . they weren't exposed too bad."

On March 12, 1981, the ACG Board reviewed the financial statements of CGD, and the directors discussed ways in which the conflict could be brought to an end. Stockton, who must have felt something like an outsider during these discussions, again spoke on behalf of CGD.

The CGD Board waited patiently to join the meeting. After the two groups convened, Mondell Mills, the CGD manager, presented a detailed, chronological review of events involving the gin's finances based on minutes and other records from the Crosbyton office. The events were discussed at length, and the two groups separated again.

Van May recalls, "Mondell . . . started as the manager of Crosbyton the same week I came on board over here [at PCCA]. Mondell is really a super manager, a very astute businessman. I think it was his business acumen that led him to discover a lot of these things that had gone undiscovered for two or three years, and he began to raise questions. As a result," May says, "I think he was kind of viewed as one of the catalysts of the problem, when in reality all he was doing was a good job as their manager, raising issues that needed to be raised. It got to the point where the Crosbyton group and the Lubbock group, through Mondell Mills and C. L. Boggs, really were beginning to have communications problems." The Crosbyton group, May says, "was making various accusations, 'We're going to sue. You people have lied to us. You have done things wrong. We haven't been compensated. We were led astray. We have everything pledged and have been given bad data.'"

Mondell Mills recalls the circumstances. He sees them in a different light than May and some of the other players. He agrees, of course, that Crosbyton had a legitimate grievance. "[We] had all [our] assets up against all the liabilities of not only the gin but also the denim mill. It couldn't have been built without those assets. . . . [We] were willing to back the mill," he says. "There was nobody [in 1974] jumping up and down and hollering and making any wild speeches about backing the denim mill." Then he downplays the conflict. He says the separation came about simply because Crosbyton "didn't want to have their gin at risk any longer than necessary," and it no longer was necessary after the denim mill began to enjoy some success. According to Mills, they just wanted to regain control of their assets.

People have wondered why the controversy arose after the denim mill began to be successful. Logic says that when times get good, people get less, not more, contentious. Mills may have answered that question in his explanation of the spin-off. And, he adds, the spin-off could not take place when times were bad, "because FmHA [and the bank] wouldn't let go of Crosbyton's assets." Crosbyton could not leave the deal until the mill was successful.

Mills also downplays the friction between the groups and individuals. The Internal Revenue Service had to rule on the tax implications of the spin-off. Mills says, "I think that probably, if you'll read into the tax ruling, there was something mentioned about friction between management. But that was Internal Revenue talk as far as I was concerned. We had to have reasons as to why this spin-off [was taking place], and I think that was probably one of the reasons [we used]." About the communications problems mentioned by Van May, Mills seems surprised. "There were some accounting problems, but I don't think there were any communications problems. The farmers in Crosbyton were getting a profit-and-loss statement but not a balance sheet, and it was difficult to really do what you think you should do without a balance sheet. I think that probably was about as much [of a] communications problem [as we had], if there *was* a communications problem." He admits to suspicion on the part of Crosbyton about the lack of a balance sheet. "There might have even been some thoughts that maybe they didn't want us to have one." He says he never believed that, only that the accounting division of PCCA was very weak before Van May got into a position where he could exercise some control over it.

As for the ACG viewpoint, Mills suspects "that there were some folks who are not involved in the [Crosbyton] Gin but are involved [in ACG] who felt the gin here was swinging along on the 'sugar tit,' when in effect we really weren't. If we were," he adds, "I wasn't aware of it. Certainly, when we spun off we owed ACG a substantial amount of money, which again indicated to me that we were not swinging along . . . but I think some of those folks thought we were. Again," he says, "if we had a balance sheet, it would have helped."

After the March 12 meeting, the ACG Board reconvened by itself. Wayne Huffaker moved and Jim Bob Curry seconded that management be authorized to negotiate a settlement with CGD, subject to approval of the bank and FmHA. The settlement proposed by Huffaker would take effect upon separation of the Crosbyton Gin Division from the rest of ACG. It would credit Crosbyton with $50,000 for the name "American Cotton Growers," which had been owned by them but would become the sole property of ACG. ACG would retain all stock owned by CGD, in the approximate amount of $1.4 million. Crosbyton would be credited with about $70,000 representing an error in recording sample rebates from the 1975 and 1976 crops, plus accumulated interest. Crosbyton also would purchase $160,000 in building bonds of ACG. Finally, all income taxes paid by ACG through September 30, 1980, would be absorbed by and remain on the books of ACG, with none appearing on the books of Crosbyton. T. W. Stockton asked whether the ACG Board had any objection to an outside auditor coming in to check the figures. No objection was stated. Six ACG board members voted in favor of Huffaker's motion, and it carried. Stockton appropriately abstained.

Wayne Huffaker played an active and vocal role in the dispute with the Crosbyton Gin Division, almost from the time he replaced Harold Barrett on the ACG Board in 1979. Van May describes him and his role. "Wayne is very outspoken, bottom line, cut through all the bull and 'let's talk eye-to-eye, nose-to-nose, if necessary' about the issues. I thought he was a tremendous board member from that standpoint, but sometimes that approach can create hard feelings on the other side, and I sure think it did in this particular case. . . . "Accusations would be made against American Cotton Growers and its board and staff, and Wayne would get very vocal in defense of [ACG] and, as a result, he became spokesman for the ACG Board position on things. The Crosbyton group reacted against that." May speculates it may have had something to do with Huffaker eventually losing his seat to Harold Barrett.

Huffaker says it probably did. "When it came time for reelection . . . [Crosbyton] had about 60 of their members over to vote against me." He explains his stand on the spin-off. "The majority of the board would have given in to most of the demands of the Crosbyton bunch, which I thought was grossly unfair to the rest of the membership financially. I thought [CGD was] demanding a whole lot more than they had coming, and we fought them tooth and nail.'' For example: ''They were demanding that they should have a substantial amount of money for the name of American Cotton Growers.'' It wasn't just that they wanted their exposure reduced, he says. It was cash, too. Overall, ''It was upwards of $300,000 that Crosbyton was demanding that they get out of the split. I ended up seeing that they didn't get that much because I didn't think they had it coming.'' But, he adds, ''Those guys were sincere. They thought they had it coming.''

Asked who his allies were on the board, he is quick to answer. ''Clarence Althof, from Roscoe turned into one of my allies pretty quick, and Jim Bob Curry became a fighter on the issue. We ended up working it out to where it worked for both sides.'' He, Althof, and Curry were on one side. Stockton was on the other. Brewer and Middlebrook were, according to Huffaker, ''pretty well in the middle,'' and Unfred, ''being chairman, was very much in the middle [and] he didn't take a stand.''

At a meeting later in the month, on March 26, L. C. Unfred asked Stockton and Mondell Mills to report for the CGD Board with respect to settlement and separation. By this point, separation was no longer just a prospect. It was going to happen. The only questions remaining were ''when'' and ''how.'' Stockton said the Crosbyton Gin Board had agreed on a list of 16 proposals for the ACG Board's consideration, and Mills talked the board through the list. The Crosbyton representatives made certain it was understood the items were ''not to be construed as everything to be considered in this transaction, but we feel these items must be addressed and agreed upon before negotiations can continue.'' Each point was discussed in detail, and some changes were made in the wording. Jim Bob Curry moved the board accept the CGD proposals for settlement as amended, subject to legal review. At the March 12 meeting, Curry had seconded Huffaker's motion. Now, Huffaker seconded Curry's. The motion carried unanimously, and the amended proposals were forwarded to the CGD Board for its consideration.

For Van May, the difficulty of working on the issue was compounded by the fact that he was a native of Crosbyton. ''Here I was, mid-twenties, young guy, politically naive as to how these things happen with farmers, and thrust into a situation of being the hometown Crosbyton boy. Either my wife or I graduated from high school with the oldest children of five of the seven men on the board over there [Crosbyton]. I had known and respected them all my life. And here I was with the ACG Board.'' May worked exclusively for ACG during his first few months on the job, until some of the accounting functions were consolidated. ''I had a great deal of loyalty to the ACG group here,'' he says, ''yet also had feelings for those people in Crosbyton.'' Eventually, his close ties to Crosbyton were a help. ''Initially, C. L. Boggs and I would go to Crosbyton board meetings,'' he recalls. ''They would start a meeting over there at about seven o'clock at night and we would sit around, and they would ask us to join them maybe at nine o'clock, and we would have accusations and discussions and present information until midnight or 1:00 a.m.

''As [the issue] progressed and it became more and more heated . . . and as the communications problems became worse between the ACG Board and the Crosbyton Board and Mondell and C. L. . . . it seemed like I was the only one left from the Lubbock operation with any credibility over there. I probably had it only because they knew me and they knew I was going to be honest about what was going on and do my best to help work it out.'' He pauses and says with some wonderment, ''With all that was riding on it, that was a pretty heavy responsibility for a 25- or 26-year-old guy.'' May describes his role. ''I would go over to Crosbyton and listen to the accusations and give my analysis to them as to what I really thought had happened and present the latest set of numbers. The numbers were always changing as we dug in and found more things. I would communicate to them what the ACG Board's feelings were, areas [in which] I felt like the board might be willing to give on an issue, and then come back and report to the ACG Board.''

As the legal and tax issues surrounding the spin-off were being raised and examined, Crosbyton could do nothing for itself. It needed to plan for its own future but was still too closely tied to ACG. Stockton told the ACG Board on April 9 that the Crosbyton Division needed to buy trucks, trailers, module builders, and accounting equipment. It also wanted to retire some of its stock. The actions would require an expenditure of about $500,000, which Crosbyton could not make without ACG Board approval. There was no action taken on the matter. It was discussed again several weeks later, and the only decision the board could reach was to postpone action once again.

It was mid-July before the Crosbyton Gin secured the equity-transfer consent agreements from its membership that were required as part of the transaction. After a great deal of time, effort, and worry, the way was clear for resolution of the conflict.

At the July, 1981, board meeting, L. C. Unfred read a letter from Billy Whorton, who asked to be replaced on the board. The letter cited his years of work on behalf of cooperatives and his desire to spend more time with his family and business. Those reasons were true enough. But, in fact, Billy Whorton also was ailing. Emphysema was slowly drawing life from him, and he could no longer maintain his position in ACG. Somber now, the members of the board recounted some of the contributions Whorton made to West Texas cooperatives, especially ACG. Then they briefly discussed the manner in which his successor would be named.

Action on replacing Whorton waited another three weeks. Then, in a telephone meeting, the results of a caucus held in District VI to fill his unexpired term were announced. Whorton had recommended that he be succeeded by Clarence Althof, who had been secretary at the Roscoe Cooperative Gin when Whorton was its president. With Whorton's support and Althof's, own good reputation, his appointment was assured. The board accepted Billy Whorton's resignation "with deep regret." It then voted unanimously to confirm the District VI caucus and appoint Althof to take his place. Althof came from a farming family. He was born in Roscoe in 1926, and after service in World War II he returned there. He was not a high-volume cotton producer, and so for 17 years he divided his time between farming, engineering a railroad train between Roscoe and Snyder, Texas, and running an insurance agency with his wife.

In October of the following year, 15 months after leaving the ACG Board, Billy Whorton died. At a meeting soon after his death, his former colleagues and friends passed a resolution honoring his leadership and dedication. They stood for a few moments in silent tribute to a man who was one of the board's original members and one of the most respected figures in West Texas cotton.

H. H. (Lindy) Lindemann, manager of the Rolling Plains Cooperative Compress, remembers Billy Whorton. "People had a lot of confidence in Billy. He didn't have to tell the people to do anything. He could just suggest it, and they would do it. People knew he had done his homework, before he brought [any idea] out in public — especially if he was trying to sell something that involved other peoples' money. If he thought it was a good idea, they knew it was. He was a real leader.'' Cooperative leaders normally devote a great deal of their personal time to their cooperative and its programs. Even by those standards, Whorton was exceptional. ''Every meeting held — annual meetings or any gin meetings that included members of ACG — Billy Whorton made it,'' says Lindemann. ''He was on the ground floor of everything that was going on. He knew everything that was happening.'' He credits Whorton with being instrumental in bringing a number of Rolling Plains gins into the ACG program, ''gins like Inadale and China Grove and Buford and Farmers Coop at Roscoe. He was the president of that board.''

Whorton, he says, was less a farmer than a businessman. ''He wasn't a man who got out in the field. He farmed a lot of land, but he did it all from behind a desk,'' while others worked the turnrows. Lindemann remembers his humorous side, too. ''When his son graduated from college with a master's degree, he said he guessed he had the only master's degree tractor driver in Roscoe, Texas. He didn't have an air-conditioned tractor cab until that boy got out of college, then he said, 'By God, I guess with a master's degree I'll have to get him an air-conditioned tractor.' And he did.''

As Clarence Althof joined the ACG Board, installation of the looms was going flawlessly in Littlefield, and the project was three weeks ahead of schedule. Completion was anticipated before the end of September. Shipments to LS&CO had averaged 601,000 yards per week since the beginning of August, some 211,000 yards above the normal flow rate, and the intense schedule had cut deeply into the inventory. Bob Hale said the loom conversion project would enjoy a cost underrun of $1 to $1.2 million. Several hundred thousand dollars of this would be due to ease of installation, and the rest to strengthening of the dollar compared with the Belgian franc.

The case of the Crosbyton spin-off was not completely closed. It reopened briefly in September, as ACG sought Internal Revenue Service (IRS) approval on some technical aspects. It necessitated a meeting in Washington, DC, between officials of the IRS and auditors and representatives from ACG and CGD. The ACG Board sent Wayne Huffaker and Doyce Middlebrook. T. W. Stockton represented the Crosbyton Gin Division. The meeting was inconclusive. The IRS wanted more information before reaching a decision which, they said, would come in 30 to 60 days.

Van May recalls the IRS approval was difficult to obtain. In trying to get the plan approved as a tax-free reorganization, ACG was setting a precedent.

"There may have been other tax-free reorganizations of cooperative divisions in the past," May says, "but I wasn't aware of any. And nobody at the Internal Revenue Service in Washington had expertise on both cooperatives and tax-free reorganizations to make this an easy transition for us." In fact, May thought they "tended to feel . . . that we were somehow trying to get away from paying tax, that we were trying to make a tax-free thing out of a transaction that should have been taxable. All we wanted to do was split up and go our separate ways." May thinks the meeting, although inconclusive, was helpful. "By the time we left, the IRS figured out we weren't trying any tax dodge, we were trying to . . . have a divorce. They ultimately approved the transaction, and we got a favorable ruling and were able to proceed." The target date for the separation agreement to expire was September 30, 1981. The ACG Board extended it for 90 days with the stipulation that, when the agreement was consummated, the effective date would be September 30, 1981.

The IRS finally responded favorably to the Crosbyton separation but said several details related to the handling of taxes were yet to be resolved. The first extension of the separation agreement would expire on December 31, 1981, two weeks after the December board meeting. More time would be needed to finalize the matter. C. L. Boggs recommended the agreement be extended for another 90 days, and it was. Except for a few technical matters, the Crosbyton spin-off was done.

Mondell Mills feels the period of the spin-off was "a good period." He says he didn't feel badly about it after it was over, because "it was just a good, hard-nosed business proposition at the time, and I think it's worked out real well for us. We got our assets." He points to ACG's subsequent success as evidence that it did not fare poorly in the spin-off. "[ACG] never did miss us. They didn't miss the assets at the time because they went from where they were just all the way up." And, contrary to the opinion that he had a rancorous relationship with C. L. Boggs, he has nothing but good remarks for the way Boggs and the ACG staff handled the spin-off.

Wayne Huffaker thinks the staff should have taken a tougher stand on the side of ACG, "which they didn't. They pretty well stayed neutral."

For Van May, who was more directly involved in the Crosbyton spin-off than anyone on the staff, the period was "the two worst years of my life." But it was not entirely negative. Despite the constant collision of logic, facts, and deep emotion involved, May admits the experience of being the go-between for the two groups wasn't all bad. He says, in fact, "It's probably one of the best things that ever happened to me, and not only because it worked out so well, since we were able to come to an agreement. It made me look good," he says frankly, "that I was able to pull it off. And it matured me in a hurry. I lost my virginity big time."

C. L. Boggs reflects on May's handling of the situation and his value to the association. "One of the best decisions I ever made was giving Van the chance. He not only did the job, he has become the best chief financial officer this cooperative ever has had." The compliment is significant. Boggs felt John Taylor was very good, and Boggs himself once served in the job.

By October, the mill was up to an unprecedented 93.5 percent efficiency. "Levi is continuing to pull out yardage at strong rate . . . 470,000 yards weekly over the past four weeks," Bob Hale said. "The plant has reached a 'levelling-off stage of operation,' which is helping efficiency. Since the rhythm of the plant can be kept at a fairly constant pitch, we are better able to coordinate all departments in a smooth running manner." The news from Littlefield was good, but storm signals were going up once again in the denim industry. Hale said the general denim market was not showing signs of weakening, but a recent American Textile Manufacturers Institute (ATMI) report showed a rise in mill-owned denim inventories over the past two years and not much reduction in production. "Something will have to give soon," he warned, "and we are beginning to see a slight amount of curtailment in the industry."

Hale's alert was followed closely by similar signals from Levi Strauss & Co. C. L. Boggs visited LS&CO and found "the present outlook . . . not as optimistic as six months earlier, since forward orders are running only a few weeks ahead and retailers are not carrying inventories." LS&CO officials told Boggs the next quarter would likely see another price decline, "since restricted exports had caused U.S. suppliers to turn to domestic markets." U.S. denim makers were trying hard to move their goods domestically at what seemed like the onset of another downturn in the market.

CGD Versus ACG

By this time, everyone on the ACG Board was becoming accustomed to the upswings and downturns of the denim market. Today Clarence Althof says, "We've been in it long enough now to know that this up-and-down is going to continue. It's been that way ever since I've been in it. We'd go from one work schedule at the plant to another and back again." He shakes his head and says, "I don't know how, but Mr. Hale could just keep his employees and continue to operate." He pauses. "It's amazing what he can do." As the latest downturn steepened, Hale told the board, "[it] reaffirms ACG's wise decision in sticking with Levi, which is feeling the effects of the economic situation but will be able to weather them better than [other makers of jeans]."

By December, 1981, weaving efficiency at the plant was at 95 percent and LS&CO quality at 97.3 percent. There was a possibility that the plant might not be able to make enough yarn to continue with weaving efficiency that high, and it could drop to around 92 percent. It would have been difficult four years earlier for anyone to believe that the plant would one day face a *drop* in efficiency to 92 percent.

There is a lag between the time economic developments indicate a downturn and the time it results in production cutbacks. The contracted flow rate to LS&CO in December was 435,000 yards per week; actual shipments were averaging 446,000. From the signs being shown in the denim marketplace, it did not appear the flow rate to LS&CO would hold for long. Once again, the issue was becoming not whether ACG could produce denim but whether it could sell it. As weeks wore on, it became apparent that the denim market was doing another of its periodic dives. Levi Strauss & Co. announced it would reduce the ACG flow rate by about 24 percent in the second quarter of 1982, and the cooperative also was told it could expect a drop in price. The board had to find alternatives to deal with the cutback while maintaining the mill's profitability and efficiency. It directed the staff to continue discussions with LS&CO to determine how long the situation might last and to see what sort of plan for dealing with it might be developed. Boggs and Hale visited LS&CO's headquarters. Then they traveled to New York to meet with sales representatives who, it appeared, might once again be needed.

Coping with Success

QUALITY UP, MARKET DOWN

March 25 was the date of the 1982 annual membership meeting. As was true throughout ACG's history, irony was the order of the day. The denim market was turning down. LS&CO had ordered a cut in the flow rate from ACG. Then, at practically the same time, LS&CO announced that ACG had achieved the ranking of "Number One" quality supplier for the quarter ending on December 31, 1981. The performance was commended in a letter from Peter Haas, Jr., who had responsibility for denim procurement at Levi Strauss & Co. With this news, the board entered the annual meeting.

C. L. Boggs reviewed ACG's uphill struggle, from its troubled beginnings to the point where it was able to meet Levi Strauss & Co.'s specifications, the toughest in the denim business. In this light, he said, "it is all the more satisfying that in the final quarter of 1981, ACG attained the ranking as Levi's number one quality supplier among all its suppliers." He told the members that denim prices had not been good for some time and they would be affected further by the recession that was slowing economic growth across the country. He acknowledged the financial pace ACG had maintained over the previous several years would be affected, but said the relationship with LS&CO and the performance of the denim mill would continue to give ACG an advantage over other textile mills.

L. C. Unfred praised the Crosbyton Gin Division, its support in originating American Cotton Growers, and its patronage over the years. He said as soon as the separation was final, the gin would be able "to perform on its own, in like manner with other cooperative gins," and wished its members well. The Crosbyton delegation was recognized by the other members of ACG with a round of applause. One of the attributes the Crosbyton Gin brought to the relationship with ACG was the quality of its cotton. It was not *always* better than other cotton in the textile pool, but often good enough to make a noticeable difference in the production of yarn. Because of the difference between producers' cotton, ACG as early as 1978 had considered the possibility of encouraging the growth of high-quality, high-strength cotton by paying more for it.

At the annual meeting, Boggs talked about plans for the premium-for-strength incentive program, which would pay premiums ranging from one to three cents per pound on 1982 crop deliveries that met certain levels of strength. The board already had voted to adopt the program. Bob Hale was pleased. After seven years, the pool was not the size required by the bank, so the mill did not yet have the luxury of taking one bale in every four from it. Because the mill had less cotton from which to choose, the quality of that cotton was of critical importance. Hale told the membership meeting that the mill was continuing to perform at an exceptional efficiency level, but the strength of the denim was only "borderline" because of uneven cotton quality.

The terms of directors in Districts III, IV, and VI were expiring. Dallas Brewer was nominated in District III, Wayne Huffaker and Harold Barrett in District IV in a replay of the 1979 board election, and Clarence Althof in District VI. There were no nominations from the floor. The results of the balloting showed the winners were Brewer, Barrett, and Althof. Harold Barrett had returned to the board, defeating the man who had turned him out three years earlier. Barrett and Huffaker never will be mistaken for each other. They both enjoy reputations as good men — well-liked and highly-respected. But they are very different. Barrett is one of the "old lions," a World War II veteran who piloted a bomber in the Pacific and is one of only two members of his flight crew still living. He went homesteading in California after the war, met and married a girl there, and then brought her back home to West Texas where he grew cotton. His eyes squint, but they sparkle between the lids. He is bright and has a wry sense of humor. He will close his eyes in the middle of a conversation, seeming to leave the discussion but actually reaching back in his mind for a recollection, an anecdote, or a humorous remark. He has many of them, and he is likely to use them to relax the tension or disarm the combatants in a conflict. After the annual meeting, the board reconvened. The directors welcomed Barrett on his return, then unanimously reelected all six association officers.

By early April, 1982, the extent of the latest denim downturn was alarmingly clear. Boggs and Hale visited again with LS&CO in San Francisco. LS&CO was anxious to resume normal flow rates as soon as the business outlook brightened. And they assured ACG they would continue to take its backlog of denim based on prior commitments. ACG would enjoy a degree of protection because of its contract. Yet little hope was given that LS&CO would purchase the excess of ACG's production from a seven-day work week.

Boggs recommended the plant be cut back to a six-day week beginning Sunday, April 11. The board was concerned about the effect of the cutback on employees, but saw no other course open to it and concurred. Several weeks later, Hale reported the six-day week at the plant had not seriously affected employee morale. He had considered recommending to the board that the plant be shut down the week following July 4, but later changed his mind, believing it best to keep running. The denim situation was such that even Levi Strauss & Co. announced it would shut down its plants for one or two weeks around the July 4 weekend.

Boggs asked Hale to contact Kurt Salmon Associates again. With the denim market so volatile, it seemed wise to take a hard look at alternatives in fabrics and markets. Boggs posed a question that answered itself in light of fluctuations in the denim market. "Should ACG research alternative manufacturing opportunities every two or three years to keep abreast of current market potentials?" Hale contacted KSA, which had begun a similar study on the same topic in February, 1980. KSA responded with a proposal in three phases: Phase 1, the feasibility of ACG producing alternative fabrics; Phase 2, an analysis of market opportunities; and Phase 3, a proposal of organizing for marketing and sales. The total study would cost ACG somewhere between $53,000 and $68,000. In mid-June, 1982, the board voted to go ahead with Phase 1.

Everywhere ACG turned, it faced unexpected costs. The Occupational Safety and Health Administration (OSHA) of the U.S. Department of Labor had taken an interest in the Littlefield plant's operations. Concerns about the effects of cotton dust inhaled by workers had turned that hazard into a major issue on Congress's agenda, and therefore it had become a priority of the agency. OSHA was the agency whose regulations made it impossible for the old Crosbyton, Wake, and McAdoo gins to remain in business nine years earlier. In that respect, the agency was partially responsible for the construction of the super-gin and the development of American Cotton Growers. Now it was examining the denim mill's compliance with dust standards. Bob Hale put his staff to work on the problem. The Littlefield plant certainly was not the only one in the textile industry under OSHA scrutiny. Other mills, particularly older ones, were hurt far more by the standards, and many were switching their production to synthetics due to the high cost of compliance.

Hale was before the board in July with a plan to meet the standards. In the preceding few months, his staff had performed tests, worked with consultants and engineers, and visited nine textile mills to obtain information they needed to develop their plan. To meet OSHA standards completely, Hale said, might take three phases of work. The work to be done was technical, requiring steps to lower the amount of cotton dust emanating from the manufacturing process and better filter dust in the plant's atmosphere. The board trusted Hale to do what was needed and do it right. But the directors were concerned about the economic impact. What price tag was attached to the work? Hale told them Phase I would cost $1,060,368, Phase II $150,000, and Phase III $850,000. He recommended they be done one at a time, because it might not be necessary to do all of them to bring the plant into compliance. The board voted unanimously to authorize the expenditures required to complete Phase I. By September, work on the dust problem was well under way, and Hale was estimating it would be completed in January, 1983.

C. L. Boggs told the board that Peter Haas, Jr., of LS&CO had called him, saying LS&CO's second quarter purchases from ACG contained an error and were underestimated by 200,000 yards, which LS&CO would take now, in the third quarter, but pay the higher, second quarter prices. It was, Boggs said, further evidence of LS&CO's integrity and sense of responsibility to ACG.

The young Haas, now director of product integrity for LS&CO, has paid close attention to the quality of all firm's suppliers, and possibly none more than ACG. The story of the cooperative's early problems and rise to prominence is one with which he is very familiar. "They will do anything and everything to try to achieve constant improvements of their fabric and maintenance of [high] standards." He acknowledges ACG is not perfect. They have "'hiccups' once in awhile, but they will work many long hours with us and on their own, as well as go outside to get any kind of technical help from anyone to correct the problem." When they have problems, Haas says, "it is not something they are defensive about or [from which they] try to absolve themselves. They will accept them, admit them, and work toward their correction." He was particularly impressed with the organized approach to quality taken by the mill under Bob Hale's leadership. He cites "the charts they have on display that monitor just every aspect of their performance, their internal measures. That is out there for all their employees to see, and I think it just another indication of their interest in constant improvement . . . and early warning and prevention of any possible problems. It is a clean plant, one of the cleanest in the industry. Conscientious people," he says of the cooperative. "Very hard working, a family atmosphere exists, and they are very, very committed to Levi Struass & Co."

SHORT COTTON

Through the spring and summer of 1982, reports came in from the field, through Jimmy Nail and Dan Martin, that several producers outside the original pool gins wanted to get into the pool but could not afford the $10-per-bale building bond requirement. The board held a number of meetings to consider the issue, but the discussion, as it so often was on the subject of the pool, was circular. It did not lead anywhere. There was much more flexibility over ginning after the 100 percent concept was put to rest. But there continued to be tremendous resistance from the board and the textile pool committee to increase the size of the pool. This was especially true where it would allow newcomers to profit from the work done and risks taken by the original members. The board was not opposed to taking in new acreage from *present* members. But the directors concluded this was not a likely course for members to take, because market conditions at the time would not allow expanded production by most of them.

All this continued to be a matter of concern to Jack Hughes, who reminded the board at every opportunity that their financing was founded on a pool size of 250,000–300,000 bales. In response, L. C. Unfred suggested that ACG might try to get FmHA to amend the loan covenants, removing the pool size requirement. The board would rather lower the requirement than expand membership to meet it. And yet, in July and August, the staff was searching for alternatives to provide the mill with 1982–1983 crop cotton.

If it were not possible to increase the overall size of the pool, the staff viewed the premium-for-strength program touted at the annual meeting as a way to increase the percentage of suitable cotton in the existing pool. The program was well-received by ACG members who invested heavily in high strength seed for the 1982 crop. But the premium-for-strength program still might not be enough to meet the mill's needs. Bad weather and the lack of available seed variety for the 1983 crop made it appear the program would suffer a setback. For the mill to have the cotton it needed, ACG might have to expand the pool. Again, the board deferred its decision. The directors decided only to continue operating "under the current pool membership policy while continuing to evaluate alternatives to increasing the pool size."

By fall, the denim downturn began to be reversed. In October, Bob Hale said new LS&CO contracts would take all the denim manufactured by the mill through the end of the year. Boggs told the board that "an upturn in the denim market appears to be on the horizon," and his staff still was looking for cotton for the mill. At that point, the staff even talked about buying cotton through TELCOT from independent producers. Boggs recognized the irony in this situation, as surely did everyone involved. The ACG Board was holding firm against expanding the pool to non-ACG cooperative members. At the same time, the staff felt it had to consider buying cotton from independents, growers who were not even part of the cooperative family.

Boggs told the board such a step clearly would conflict with PCCA's long-standing policy of marketing only cooperatively ginned and stored cotton, but it would "have to be studied further." One month later, the staff recommended ACG not pursue that course. It appeared there would be enough cotton from the cooperative sector to run the textile mill. It was good that there would be. The downturn of a few months earlier had been so completely reversed that Levi Strauss & Co. was requesting more ACG denim "if possible." No one but the worst cynic would suggest the staff's search for independent cotton was aimed at moving the board to a decision on expansion of the pool within the cooperative community. But it certainly was something for the directors to keep in mind as they deliberated about how they would obtain cotton to run the mill.

The 1982 year-end financial results, the first full year since Crosbyton had separated from ACG, showed net taxable income of $3.7 million. Van May said the association's working capital at the end of the year was $9.5 million, with a debt to equity ratio of 0.83 to 1. Both figures exceeded the bank's requirements. It appeared Mondell Mills was right; the association came through its first year since the spin-off with barely a hitch.

GREETINGS FROM LITTLEFIELD

New Year greetings in 1983 came in the form of a letter that ACG and 16 other businesses received from the city of Littlefield. The letter said that the businesses were going to be annexed by the city, placed within the city limits, and subject to payment of municipal taxes. The ACG staff joined the other businesses and legal counsel in a meeting that ended with a decision to draft a petition to be signed by all the companies in opposition to the measure. The city had not provided services to the businesses, nor was it likely that it could. Why should they pay? The petition would suggest the city discontinue its plans. The staff could not sign the petition without the board's approval. At a combined February-March meeting, the board gave its unanimous consent. It authorized Emerson Tucker to sign the petition, and Tucker and Bob Hale to negotiate with the city.

In this case, the word "negotiate" has to be taken advisedly. Negotiating with a governmental jurisdiction that has the power of eminent domain is sort of like negotiating with someone who is armed. The compromise tends to be one-sided. Talks were ongoing between the businesses and the city, but it was not until August, seven months after the city sent the letter, that there was real movement on the issue. Hale and some of the business owners met with Littlefield's mayor and city council. The council had voted against annexation but wanted to continue discussions with the property owners. Hale reported to the ACG Board that he believed the issue would be resolved — at least insofar as ACG was concerned — if the cooperative agreed to pay a "reduced taxation" rate of $70,000 to $80,000 for the next five years as an alternative to annexation.

A month later, the city council showed its vote against annexation was meaningless. Instead of annexing the mill property outright, it presented ACG with a choice. The cooperative could pay $80,000 per year or see its mill annexed into the city limits of Littlefield. It was a "Heads we win, tails you lose" proposition. After further talks, the two parties agreed ACG would pay $75,000 per year for seven years, with no rate increases, in lieu of its property being annexed. The ACG Board, having little choice in the matter, approved the settlement. It would cost over one-half million dollars but was less expensive than the alternative. Van May estimates that annexation would have cost ACG about twice that much in taxes.

In March, 1983, the issue of pool membership again was on the agenda. The board decided to accept membership applications from growers at non-pool gins, but only under certain conditions. These were included in a motion put forward by T. W. Stockton, seconded by Jim Bob Curry, and approved unanimously. The production of the farms requesting membership from the gin must have equaled fifty percent or more of the gin's last five-year average ginning volume. And the gin still would be required to purchase building bonds equal to $10 per bale times the past five-year average ginning volume from the farms entering the program. The conditions, which would prove difficult to meet, would satisfy ACG that non-pool gins entering the program would do so wholeheartedly, with a majority of their ginnings, and that their financial commitment would be sufficient to earn them a place in the program.

The 1983 annual meeting was held on March 24, and the reports were exhilarating. All the key numbers — denim prices, profit margins from the mill, members' equity versus long-term debt, cash distribution to members versus average TELCOT sales, profitability of the new looms, the revolving capital credit plan — were good and steadily getting better. There was little or nothing about which anyone could be displeased.

One month later, the mill achieved a weave room efficiency of 97.5 percent for the first time in its history. And tests at Texas Tech concluded that the mill now was meeting OSHA dust standards.

The opening of the pool, despite its stringent conditions, began to show results. The staff's estimate of 1983 participation showed an increase of 117,684 acres from new sign-ins. For the first time, the pool was projected to meet, even exceed, the goal of 300,000 bales. The estimate included totals from members applying, but not yet accepted, from gins at Midkiff, Fieldton, and Spade.

SHUTTING THE POOL

The board then shut off the flow into the pool, limiting deliveries to those gins currently in the program. In acting unilaterally, the board lost sight of the advice the staff had given it in 1978, when the

100 percent concept was put to rest. Any decision on the pool, the staff had suggested, should include the ideas and opinions of the non-pool cooperative growers and gins. Once the pool reached its 300,000-bale limit, the directors said they would accept only those applications received from gins in ACG's Lubbock office or postmarked before 5:00 p.m. on Friday, April 22, 1983. This action was taken on April 21, so the chance of new applications coming in by the deadline was remote. All applications received after that date would be placed on a waiting list in order of the date received. The board then approved pending applications, as well as textile pool committee representation, for the new pool gins and assigned them to districts. Fieldton and Spade would go to District II, Midkiff to District VI.

By May, the pool was over 400,000 acres, with projected yields of 335,000 bales. C. L. Boggs said the figures were slightly overstated and the actual bales would be about 300,000 when the acreage was returned to full cotton production. The numbers for the 1983 crop would not look as good. Because of government set-aside and other PIK (Payment-In-Kind) program participation, the 1983 crop would be only about 175,000 bales.

At the end of May, Jack Hughes was keynote speaker at a joint meeting of the ACG Board and the executive committees of the regional cooperatives. He used the occasion to explain the scope and purpose of a new study of cooperatives in the Rolling Plains and High Plains areas of West Texas. The objective, he said, was to identify how the cooperatives were changing, how they most likely would change in the future, and what initiatives industry leaders would have to take to be responsive to their members' needs. Darryl Lindsey, ACG vice-president for operations, says that some observers interpreted this study as a basis for putting the regionals together. He probably is right. As early as January, 1980, Jack Hughes was advocating the "advantages of [the regional] cooperatives combining their assets, and the benefits that this integration would afford" in remarks to the ACG Board. Dr. James Haskell, then director of the Cooperative Marketing and Purchasing Division of USDA, had been asked to coordinate the study. Hughes introduced him to the joint meeting. His goal, Haskell said, was to illustrate objectively any economic benefits to cotton producers that would result from a more efficient cotton-handling system in the High and Rolling Plains. The scope of the study would include the entire area cotton industry, from the turnrow to the textile mills.

By October, 1983, the denim market began to fall again, this one more a slide than a dive. In November and December, inventories at Littlefield grew as LS&CO shipments slowed. Bob Hale said the mill, which had been operating all-out for almost a year, would close for 10 days during the Christmas and New Year holiday period and would resume with a six-day work week beginning January 1. He also was considering curtailment of production during the first or second quarter of 1984, depending on LS&CO's needs.

The economic squeeze carried into 1984. At a joint meeting of the PCCA Executive Committee and ACG Board, Boggs asked that no salary adjustments for the staff be recommended for the year because of the "tight economic conditions facing the cooperative's members." But financial issues were not to be the major challenge facing the cooperative in 1984. A serious competitive situation was building between the pool gins and the non-pool gins. Even though restrictions on where individual producers could gin had been liberalized years earlier, Boggs says, "We continued the concept all the way through, that the only way you could get into the American Cotton Growers program, the only way you could share in the profits of the denim mill, was to gin your cotton at one of the gins that was in the program."

Nothing brings more attention than conspicuous success. As ACG started paying big cash dividends, the pool gins started drawing cotton away from the gins that were not in the program. "As manager of both regional coops, ACG and PCCA," Boggs says, "I was in a very difficult spot, and so was my staff. We were managing two coops. One had become very successful, paying big cash dividends. And it was taking business away from the other group of gins. And yet the board of directors of ACG did not see fit to change the rules. It was creating a very emotional and difficult situation."

When the board needed to build the pool, it took steps to liberalize membership requirements. When the pool reached the right size, the board began to shut the flow into it. Now it sought to restrict the pool even further, including the extent to which present members could participate. A February 16 joint meeting of the ACG Board and textile pool committee discussed the issue. Two major questions drove the talks. First, did the cooperative want to continue the policy that permitted unlimited expansion of cotton acreage on farms currently in the pool? Second, should the membership right be attached to the farm or to the individual landlord or tenant? The group could not reach an answer to either question.

Wayne Huffaker, then serving on the textile pool committee, suggested membership rights be established on the cotton base acres then in the pool. That acreage would form the foundation for any future adjustments. He also said membership rights should be with the individual rather than the land; that rights should be salable, transferable, assignable, or inheritable within each gin's territory; and that if the rights were not used in two years, they should revert back to ACG at no cost. As an alternative, he said, ACG could establish a price it would be willing to pay to purchase the rights. Some of the board members voiced concern. They feared if membership rights were with the individual rather than attached to the land, with only part of a farm's production in the pool, it would give rise to administrative problems. As in the past, they were opposed to any expansion in the size of the pool because of the dilution of earnings they feared from accepting additional acres or bales.

Several at the meeting suggested ACG continue the pool as it was for one more year . . . thus accepting into the pool additional acres on a current "pool" farm, and allowing current members to transfer acres if needed. They pointed out that some producers already had made financial plans for the 1984 crop that included adding additional acres to a farm already in the pool. Others objected. They said the acres should be "frozen" now and additional acres not allowed in the pool. They discussed an option of limiting the size of the pool sharing in the textile mill margins, with excess bales marketed in a separate cotton marketing pool, a so-called "two-pool" option. But they took no action.

On March 22, before going into the annual membership meeting for 1984, the board denied a request from a non-pool gin to gin ACG members' cotton. It confirmed its policy of allowing only current pool gins to gin pool cotton, and excluding from the pool individual growers not then ginning at a pool gin. Non-pool gins, the board reiterated, would be considered for membership *only* when 50 percent or more production of the five-year average ginning volume was committed to ACG and the requisite amount of building bonds was purchased.

C. L. Boggs raised the two-pool option first discussed in February. He did not favor that option. The staff felt such a plan would be difficult to administer and could lead to member dissatisfaction. Nonetheless, wanting to provide the board with analysis of all the avenues open to it, the staff prepared a memorandum showing how the system might work if implemented. Essentially, two pools would be established for sharing in the margins of ACG. For ease of identification, they were called the mill pool and the lint pool. The mill pool would share in both the lint proceeds and the textile mill margins. The lint pool would share only in the lint proceeds.

Then a second memorandum was distributed. It had been prepared by Van May and was entitled "Limiting ACG Pool Size." It showed the type of thoughtful, persuasive analysis that typified the ACG staff. It also was a perfect example of seeming to embrace an argument while actually turning it around. May began by agreeing with the basic premise of the board. "The fundamental question is, 'Should the pool be limited at all?'" he asked rhetorically. "Clearly, the answer is yes." Having said this, he went on to show a clear case *against* further limitations in pool size. He complimented the board's action of the previous year, ". . . which limited deliveries to the pool to only the current gins in the program." In his opinion, it had "constricted the pool size to a manageable level." Turning this praise against the anti-expansionist position, he concluded, "Therefore, additional limits on pool size should *not* be necessary." He wrote that restrictions on sources of cotton were stunting economic progress. "The potential for substantial growth is limited due to the limitation of accepting bales from pool gins only.

"During the . . . '100% concept' . . . crop years 1975 to 1977, the pool averaged 266,000 bales per year. [By] the addition of those who have joined the program since 1977, we might add another 20,000 bales making the average pool size of new . . . and old gins [when they were 100 percent committed] in the range of 286,000 bales per year. Obviously, some years would have been much higher and others much lower.

"However, the 100% concept is gone, therefore, unless some of the original gins experienced a tremendous growth [which they did later as a result of getting acreage from non-pool gins], the maximum potential size of the pool should not be unwieldy even if everyone at those gin communities signed back into the program."

Addressing a major concern of the board, he attacked the issue of dilution. "The big dilution of mill earnings," he wrote, "occurred last April. During the . . . 1978 through the 1982 crops, the time after which the 100% concept was abandoned, the pool has averaged 130,682 bales per year [and] the average mill earnings per crop year were approximately $10.2 million. This equated to . . . earnings per bale of $78.28 for the average of the five crops.

Had the pool size been 300,000 bales during this period, the average earnings per bale from mill margins would have been $34.10.

"This demonstrates that the sign-ins last April [to bring the pool up to the 300,000 bale level] diluted the mill margins per bale by 56% ($44.18). Should the pool go as high as 400,000 bales, the $34.10 level would only drop about $8.50 a bale, down to the $25.00 range. I think this dilution is fairly insignificant when compared to the dilution which has already taken place." One of the board's major arguments over the years for limiting pool size was fairness to those who had been in the pool from the beginning. It always was a major concern. "During the discussions about limiting the pool size," May wrote, "there's been a lot of talk about being 'fair' to our old members who have been with us all along. I think what has gone unmentioned is the 'fairness' that has always been built into our Capital Revolving Plan.

"As long as the $300 per bale limit is not raised, the members who stayed with us through the hard times will benefit by having their capital requirement filled first, and thereby beginning to receive in cash all of the earnings which [accrue] to them. . . . I think this plan is specifically designed to promote the kind of fairness that has been discussed of late.

"In my opinion," he continued, "it is a much more appropriate vehicle for demonstrating the value of having been in the program all along, as it already has a dollar value assigned to it. Additionally, the people who have been with us all along will probably have filled their capital requirements much quicker than future members will be able to do so, assuming that the pool does not diminish again."

Finally, he tried to drive a nail into the coffin of the "two-pool" option. "Administratively," he said, "we need 100% of a farm in or out of the program to properly manage the delivery process. Current discussions appear to favor having 'split farms' in ACG. From a manageability standpoint, this could become a nightmare. I think the worst problem could be created at the gin level and create a lot of unrest and dissension toward us. I think auditing the deliveries to the pool under this 'split farm' program would be impractical, if not impossible." With May's memorandum in hand, the board and staff went into the annual meeting.

L. C. Unfred introduced the board and reviewed the major accomplishments of the past year, among them, "building the pool to the optimum size and then closing it while working on a policy to equitably admit new members into the program." Among the achievements since the last time the members had convened were: record denim mill margins; record cash distribution to ACG members as compared to non-pool sales; the first retirement of capital credits issued from the ACG pool; record high denim plant efficiency (96.9 percent); record high production of LS&CO quality (98 percent); record number of quarters as LS&CO's number one supplier (all four quarters of the past year); record number of bales consumed and number of yards produced at the denim mill (86,000 and 26 million respectively); and record premium-for-strength payment ($1 million).

It was hard to believe that, just 10 years earlier, so many experts were saying it couldn't be done, that high-quality denim could not be made using open-end spun West Texas cotton. If ever an organization had cause to celebrate, it was ACG. Appropriate to the occasion was the keynote speaker, Robert D. Haas, recently named president-elect of Levi Strauss & Co., and great-great-grandnephew of founder Levi Strauss. Haas admits to having been among the skeptics years earlier. "It was a relatively new technology, and other mills that had tried it in other parts of the world had not come up with satisfactory quality denim." He says he now knows the answer. "Obviously, they did not have the benefit of West Texas cotton, which is a critical ingredient to the success of the Littlefield mill." He had paid a brief visit to Littlefield in 1981 and remembers being "impressed with . . . the efficiency and cleanliness [of the plant] and dedication of the people." On this visit, he would have an opportunity to get much closer to them. Haas, a very warm and articulate man, recalls the experience. "The evening of my arrival, before I was to address the members, I went out to dinner with the board of directors. What impressed me was, although this was a group of people that I had not met before, [and] they were clearly in a different profession than my own, they were the kind of people with whom Levi Strauss himself would have wanted to make a handshake agreement. They are straight-talking, sound-thinking, warm individuals from very personal backgrounds, with a great deal of pride in their work as cotton farmers and their affiliation with ACG and the mill they have created." He calls the meeting with them, ". . . one of the memorable times, not just in my association with ACG but all my business associations."

He reported to the membership meeting that LS&CO's share of the denim market continued to look good, although he acknowledges now that, "Levi's was going through some difficult times, and I'm sure my attempts to reassure the shareholders of ACG may have seemed a bit shaky to them." To take advantage of its prospects, the company had become a sponsor of the 1984 summer Olympic games. It also had taken steps to diversify, a refocusing of the firm that resulted in what Haas now calls, "a much sounder future." He says today, "The opportunity to address the [members] was, to me, a lot of fun and also very gratifying because I happen to believe in employee-owned enterprises or investor-owned enterprises."

Elections were held for directors from Districts V and VII. Unfred and Stockton, who had painstakingly guided the organization through the worst of times, were returned to office by acclamation in these, which seemed the best of times. Before them, like a wall, still stood the critical issue of pool membership.

On the day after the annual meeting, the board met with the textile pool committee. Speaking in support of Van May's plan, Boggs said it would allow new members to enter the pool at current pool gins only. The pool gins would be able to increase their volume of ginning through the added attraction to producers of sharing in the mill margins.

On April 4, the textile pool committee recommended the ACG Board reopen the pool to growers who ginned their cotton at the current 27 pool gins. The board acquiesced. In a special meeting that same afternoon, the directors unanimously approved the following admission policy: "The pool is reopened for membership to all growers who gin at any of the 27 pool gins currently in the program. Growers will be able to sign into the program during the month of April, which is the traditional sign-in, sign-out month. Growers currently in the pool who gin at non-pool gins will remain in ACG but will not be allowed to enter additional farms, unless the new farm production is ginned at one of the 27 pool gins."

ACG had settled on a membership policy. It would allow expansion in the pool, but only through growers who ginned at pool gins. One problem was solved but a more serious one created, one Van May today calls "the genesis of the ACG – PCCA conflict. . . . The PCCA people said, 'We helped ACG get started, all the regionals put in building bonds, we provided staff, we have done everything . . . and now you are saying that someone who gins at my gin can't be a part of that program, even if they are willing to meet all the other requirements?' Of course," May adds, "in the meantime, ACG had become very profitable, very strong, people were getting a lot of money out of it, and they didn't want it any bigger.'' The stage was set for a three-year confrontation with the non-pool cooperative gins.

BITTER PARTING

At the April 16, 1984, ACG board meeting, Boggs told the directors what he had told the PCCA Board the day before. The agreement between PCCA and CXS would be terminated. Bert Kyle and he had visited TELCOT buyers in Memphis, Dallas, and Lubbock to explain PCCA's decision to end the contract for sharing TELCOT facilities and personnel. The termination would have been effective on June 30, but CXS later chose to exercise a 12-month extension provision in the contract. ''Dan's company,'' Boggs recalls, ''did not prosper as he envisioned it would. The costs we were sharing became rather controversial. He began to pick at things . . . thought we were allocating the costs incorrectly. We were not. We were doing it exactly as we had set out to do it.'' He ascribes a rationale to Davis's attitude. ''Of course, when you have financial difficulties, you begin to question things that you otherwise would not.''

The falling-out between PCCA and CXS began to take shape months earlier. It began as a simple business disagreement. Before it was over, it would run much deeper and leave wounds that took years to heal. In Boggs's words, it ''was probably the real fork in the road as far as the relationship between Dan and me.'' He explains, ''In late 1983, [Dan] came up with the idea that we could lower the cost of TELCOT if we'd put personal computers out in the offices of the cotton gins and off-load some of the work that was presently being done on the host computer to the satellites in the gin offices.''

Dan Martin says, ''He wanted to force us to change our equipment in the country. He wanted to go with PCs [personal computers] instead of the dummy terminals that we had sitting out there. And he wanted to force the issue.''

Boggs says he was not warm to the idea. ''First, I thought it was premature. I thought, 'Again, Dan is four or five years ahead of his time,' and I think it's proved to be true.'' PCCA eventually did convert to personal computers in gins and buyers' offices, but not until 1989. ''The second reason I didn't want to do it,'' Boggs adds, ''is because we were in such a gosh-awful mess as far as internal programming, and so forth. We just didn't have the staff time to do it.'' He says he suggested to Davis that it was ''something we ought to postpone for 12 months, study it further, give us time to finish our internal programming.'' Davis was not happy with the decision, and apparently he was determined to go ahead whether PCCA went along or not. On April 2, 1984, came the event that would be a turning point in the relationship between the former manager and his understudy.

''[Dan] called and said he wanted to meet with me,'' Boggs recalls. It was not unusual; the two met regularly. ''He came out to my office, walked in, and closed the door. I think his opening comments were that four of our top programmers were coming to work for him.'' The news rocked Boggs. It was completely unexpected. What made matters even worse, says Boggs, was that one of the programmers had been ''my second closest personal friend in the world. I don't know how to describe how I felt.'' According to Boggs, Davis never made known his plans to employ any of PCCA's staff. He arranged it all, then presented it to Boggs as a ''fait accompli.'' The four employees were giving PCCA 30 day's notice. And although PCCA had about 20 programmers on its rolls at the time, these were four of the best. They were, according to Boggs, ''. . . the people who were in on the beginning . . . of TELCOT and writing it. They were the guys who knew the TELCOT system. They were what we called our 'long-range development group.''' After conferring with his staff, Boggs took immediate action. He authorized the shift of responsibilities being carried by the four over to other programmers. And he ''decided the thing we ought to do was to get those guys off our premises, period, because they had access to the 'guts' of our company. They were tied into our computer via direct line. We realized,'' he says, ''that if those guys wanted to really sabotage us or do us any harm, how much damage they could do.'' He quickly adds, ''I didn't really think they would do that. I don't think even today that they would have done that. . . .'' He pauses and, justifying his reaction at the time, says quietly, ''But I didn't think that they would all resign and send word to me the way that they did, either. After that happened, I didn't know what might happen next.''

Boggs decided, too, after conferring with his vice-presidents — May, Lindsey, and Kyle — to terminate the agreement between PCCA and CXS. The decision was neither easy nor quick. Dan Martin recalls, ''We had gone to Altus, Oklahoma, for [a meeting] at the compress there. Coming back on the plane was when [C. L.] made the decision to notify Dan the next day that PCCA was terminating his contract." The cancellation would be subject to board approval. Martin says "Van [May] and Darryl [Lindsey] had been hammering at C. L. for probably two weeks to do it, and he kept resisting." He adds that they didn't know Dan Davis the way he and Boggs did, and they probably couldn't understand why Boggs seemed so reluctant to cut him loose. "C. L. finally made the decision." If what Davis had done moved Boggs to disappointment, what he had to do in return moved him to despair.

April 18 was PCCA's regularly scheduled board meeting and one of the days Boggs always will remember. At the conclusion of the meeting, he notified the board of Davis's having hired the four programmers. He recommended that PCCA's contract with CXS be terminated. The board quickly approved his recommendation. Boggs recalls, "I told the board I was leaving right then to go to 40th and Avenue A, a satellite office we rented several blocks away, to ask those four programmers to leave the premises that afternoon and to ask for their keys to the office." He took Darryl Lindsey with him. When they arrived, Boggs collected the keys and "asked them to collect their personal effects and be out of the office within two hours. It was a very difficult thing for me to do, and I think those people feel, even today, that I was retaliating. I wasn't retaliating. I just had this company that I had to protect and I felt that [in light of] the events that had happened . . . that's what needed to be done."

There was a 12-month cancellation provision in the agreement. Looking back to 1978 when the agreement was drawn up, Boggs says Davis "asked for the cancellation provision in order to protect himself, which I think was reasonable. He was going to build a business completely tied to our computer. The software to drive that computer belonged to us. As part of the agreement to share this, he was going to, over a five-year period, own a half-interest in that software. But for him to disconnect or be disconnected from our system, and have to bring up a new one from scratch and get it going, would probably take the good part of 12 months." Boggs says before the incident involving the programmers, PCCA was "willing to work with [Davis, despite the fact that] we'd been having controversy . . . on the cost allocations . . . and it was taking a lot of staff time. . . ." The hiring of the programmers triggered the decision to terminate the contract. Nonetheless, in an offer that seems remarkable in view of events, Boggs says, "we told him that we'd honor the 12-month cancellation provision, but that if he wanted to terminate earlier he could. I think they first said that they wanted to terminate earlier because they felt like they could get their system up and going. Later they changed their minds and asked us to honor the full 12 months, and we agreed to do it."

This incident may show more clearly than any other the familial nature of the cooperatives. Only in families are such grievances, either actual or perceived, so delicately handled. It is hard to see most business organizations responding to such an event with anything short of legal action to terminate the agreement immediately and seek damages. As to what it was like to continue the agreement in light of what happened, Boggs says, "It's a long time when you want to get disconnected from someone." In March, 1985, 11 months after the notice of termination was issued, CXS filed for protection under Chapter 11. In speaking of the episode, Boggs admits, "I've never had anything in my life that was so disappointing and was such a shock to me. It probably was one of the most difficult things I've ever handled in my life."

Dan Martin confirms that terminating the agreement with Davis probably was "the hardest thing" C. L. Boggs ever had to do. "The denim mill wouldn't be there if it weren't for Dan Davis," Martin says, "there wouldn't be a TELCOT except for Dan Davis. They were his ideas. They were born in his brain." He provides a personal view of Dan Davis. "After I had my second heart surgery, I almost died. I spent 28 days in the hospital. There was not a single day that went by," he says, the emotion showing in his voice, "that Dan Davis was not at the hospital, either in the morning or at night, checking on me and carrying my wife either to breakfast or dinner. He sat with my family during both my heart surgeries, four hours the first time and six hours the second. How do you think I feel about Dan Davis?"

As for his relationship with Davis, Boggs says the two have not had much contact since that day in his office in 1984. "We see each other socially, and occasionally we will see each other at business events. We are cordial and friendly. I don't have any animosity toward Dan. I wish him the best." The tone of voice and expression accompanying the words seem based on 20 years of admiration, not on the bitterness left by one unfortunate incident.

In the Spring of 1984, the denim industry still had not rebounded from the downturn of late 1983. The media were reporting on plant closings at Levi Strauss & Co. The staff monitored the situation for ACG and said LS&CO officials were correcting errors they had made earlier in overexpanding their production capacity. For many companies, plant closings are not unusual. For Levi Strauss & Co., they must have been a bitter medicine, totally out of line with the company's record of success and its well-known concern for employees. In retrospect, Levi Strauss & Co. had not overgrown; the world denim market had undergrown.

By August, there were continuing cutbacks in denim orders, and the inventory was expected to be up to 1.5 million yards on September 30. LS&CO shipments increased following July 4, but had slowed more recently, and it appeared the reduction in flow rate might necessitate cutbacks at the mill. Levi Strauss & Co. was not the only jeans producer facing cutbacks. Blue Bell was closing its plants in Puerto Rico and consolidating those functions with U.S. plants.

Boggs had been expecting a sharp reduction in the flow rate but was surprised at the sharp drop in denim prices. He cautioned the board that the mill's profit margins for the next few months would be slight if at all. Bert Kyle said cotton prices were down because of slow textile demand, good world prospects for cotton production, and the prospect of a large Chinese crop. Bert Kyle is PCCA's "eyes and ears" in world cotton markets. Charged with one of the most difficult responsibilities in the cooperative, he has to track world cotton developments and make buying and selling decisions in a highly volatile and unpredictable arena. A native of Memphis and a World War II naval aviator, he came from a cotton family and has been in the business all his life. He has worked for PCCA since 1963, and Boggs thinks highly of him. "He has done a tremendous job. In my opinion, he is one of the best cotton men there is."

Following ACG's pool membership policy over the years is like traveling through a labyrinth. It is easy to get lost in the board's deliberations, and anyone who does can take heart from director Dallas Brewer. An erudite gentleman who now serves as a county judge, Brewer was heard to seek clarification from his fellow board members and the staff at the July, 1984, meeting. The subject was non-pool gins ginning pool cotton. In frustration, he asked, "Just what *is* our policy?"

Boggs explained, in sum, that under the current policy only pool gins were allowed to gin pool cotton. Determining the fine points was not quite as easy. All through the rest of the year, Dan Martin brought requests before the board for sign-ins to the pool and sign-outs from the pool. Each had to be judged on its own merits.

Having developed its pool policy in April, the board then had to answer questions and make decisions on cases arising from its implementation.

Windham Cooperative Gin, one of the original pool gins, was discussing a possible merger with Terry County Cooperative Gin or Loop Cooperative Gin. They needed the board's guidance on how such a merger might affect the status of their members. The staff recommended the current pool members at Windham be allowed to stay in the pool if they merge with another gin, and the successor gin be permitted to have a representative on the textile pool committee. It won unanimous board approval.

Some of the Rolling Plains pool gins had met with Jimmy Nail, Dan Martin, and H. H. Lindemann, manager of the Rolling Plains Cooperative Compress. Some of them would not have enough cotton to justify opening. Their only alternative to not ginning at all was to contract with non-pool or independent gins. Boggs brought the issue to the board, telling the directors there was some dissension among gins on how and where the available cotton would be ginned. This cautionary note underscored the seriousness of the conflict that was building between the pool and non-pool gins. These gentle winds of dissent soon would become a hurricane. Boggs said the pool gins were seeking guidelines on how to handle the situation. As far as the staff was concerned, he added, it did not make a lot of difference where the cotton actually was ginned, as long as it was tagged with the pool gin's bale tags and the bales were stored at the cooperative compress. The board decided it should not tell any gin where to gin its cotton but instead establish "general policy guidelines" for the pool gins to use in making their own decisions on where they gin.

Times were difficult on the Rolling Plains. Another pool gin from the area sought help from ACG, but it was beyond the association's ability to provide. The financial situation at the Roscoe Farmers Cooperative Gin was perilous. J. B. Cooper, the gin's president, explained to the board that the producers at the gin had suffered three consecutive crop failures and they were looking for financing alternatives. The possibility was raised of ACG lending the members of the Roscoe Gin funds to be invested in the gin's preferred stock, with the members pledging their capital credits in ACG as collateral. Bill Blackledge advised the board against it. He was concerned about prior lienholders on capital credits unknown to ACG. He also said if ACG offered financing to Roscoe producers that other ACG members might request similar treatment. After a long discussion filled with anxiety over the plight of the Roscoe Gin and concern expressed for its members, Harold Barrett moved and T. W. Stockton seconded that ACG not be involved in providing financing for the gin or its members. The motion carried unanimously.

One of the votes was cast by Clarence Althof, the ACG director whose home gin is Roscoe. For him, the vote was a responsible and conscientious act for the greater good, and it required a lot of personal courage. It was hard. But he understood why the vote had to be as it was. He expressed his appreciation for the board's willingness to consider financing for Roscoe, and told them the gin's executive committee had met with TBC officials and prepared a financing proposal for the bank to consider. There was a happy postscript to the story. After balancing on the brink of financial disaster, the Roscoe Gin eventually found the support it needed. That, and a few good crop years, returned the gin to financial health.

In September, Bob Hale said the staff should be prepared for changeover to a five-day, three-shift operation at the denim mill. Due to the decreased schedule, 1983 crop cotton would be consumed at a slower rate, and attempts would be made to sell excess bales not needed by the mill. He also warned that textile imports to the U.S. were continuing to rise. The latest downturn in the denim market could have hurt the Littlefield operation far worse than it did. By October, weaving efficiency was up to 98 percent, the best in the mill's history. Employees had adjusted to the five-day schedule, though they were not pleased with the change. The mill would shut down for two weeks in the fourth quarter to avoid a buildup of inventory. LS&CO was behind on pulling out some of its fourth quarter contract but was continuing to take the denim and honoring old contract prices. Hale said the staff was continuing to explore other sales opportunities, including foreign markets. The denim market still was flat, he said, but "bright spots are being recognized."

At year's end, denim shipments had picked up, and it was estimated that shipments for the month of December would reach 1.5 million yards. LS&CO was behind on its flow rate but there had been good followup on old orders. Projected flow rate for the mill in the first quarter of the new year was up 10 percent, and inventories on hand would be covered by orders.

The first mix of new-crop (1984-1985) cotton appeared to be dirty and long with low micronaire but reflecting good strength.

Hale said the mill would use special cleaning procedures for the worst cotton. The low mic cotton would be hard to clean but should produce good denim.

Cotton strength was an ongoing concern of the mill and the board. In January, 1985, the directors asked Bert Kyle about the future of the premium-for-strength program, because many members needed to begin purchasing planting seed for the next crop. He said if ACG wanted to encourage members to grow strong cotton for the denim mill, then premium-for-strength payments should be continued. But, Kyle cautioned, current market conditions were not reflecting much of a premium-for-strength. ACG might pay its members more for high quality cotton, but it was questionable whether they would realize an added margin in selling what the mill did not use. The program had been a good one, in Kyle's opinion, but the board might want to consider restricting future premiums to certain qualities. For his part, Bob Hale was an unrelenting advocate for the strongest possible cotton. He told the board ACG must never lose sight of how important strength is to the operation of the textile mill. L. C. Unfred told the staff to announce in the ACG newsletter that the premium-for-strength program would continue for the 1985 crop and details would be provided later.

Bert Kyle has an interesting perspective on cotton quality, and it is not the same as Hale's. All along, Kyle's goal was to turn over to the mill "those bales that were the least in value in the commercial world but that were still qualified to make denim [to LS&CO specifications]. The idea was that we wanted to make denim out of the cheapest end of our cotton that we could, so that the remainder of the cotton could be sold on the commercial market at a much higher price. . . . We produce a lot of cotton out here in most any year that is too good to be used in denim from the standpoint of cost. You would rather use something that is cheaper that will do the job. There's no need in spending 60-cent [per pound] cotton to make a three dollar yard of denim when you can make it just as handily out of 50-cent cotton. So," he adds, "we use the 50-cent cotton to make denim, and we sell the 60-cent cotton to somebody who is making something that requires a higher quality and a more valuable piece of cotton." Asked about whether this puts him in conflict with Bob Hale, Kyle says there is no problem. He understands and appreciates the need to meet LS&CO standards. With a grin, however, he adds, "Bob always kids us about just barely giving him good enough cotton."

By February, the denim downturn of mid- to late-1984 had reversed itself. The roller coaster was going back up. Textile imports continued to be a concern, but denim mills in the U.S. were beginning to step up production due to increased orders and shipments. Hale said if current shipping levels were maintained, Levi-quality inventory would be completely exhausted within five weeks. As a result of this projection, the mill would operate two Saturdays to build a working inventory and to cover orders. Demand had improved to the extent that LS&CO was beginning to use contract cutters, in addition to their own cutters, to keep up with consumer demand.

On March 21, Hale told the board the plant would change to a seven-day, four-shift production schedule at the end of the month. One hundred and five employees would be added to handle the increased schedule. The flow rate was up to 500,000 yards per week. Despite lower denim prices, the mill was projecting net margins of about $4.5 million for the current fiscal year.

The annual meeting was held that day. The members were told that ACG had recorded its second best year ever in 1984, despite problems in the domestic textile industry. Directors Brewer, Barrett, and Althof were unopposed and reelected by acclamation. In a change of officers, the board elected John Johnson as assistant secretary, replacing Mac Cooper who had resigned. Out of concern over agricultural legislation and textile imports, the board also voted to retain Vern Highley to represent the association in Washington, DC. Highley, who had left the cooperative in 1981 to take a political appointment in the U.S. Department of Agriculture, had left government and offered his services to ACG as a consultant.

Trouble in the Family III
ACG Versus PCCA

STORM CLOUDS ON THE PLAINS

Since the board's last edict on pool membership, the staff had heard repeated complaints from the members of the non-pool gins. What began as a subtle rumble of disgruntlement was increasing in pitch and volume. On April 15, C. L. Boggs reported to the board on a PCCA Executive Committee meeting in March at which Jim Bob Porterfield discussed ACG's policy of accepting cotton only from pool gins. The executive committee did not explore the issue in depth because of time constraints, but they placed it on the agenda for their April 17 meeting. Boggs repeated a suggestion originally made by L. C. Unfred, that it "might be a good public relations move" to invite the executive committee to join the ACG Board for lunch on April 18. The directors agreed.

The board convened again just prior to its meeting with the PCCA Executive Committee. Unfred said it might again be time to review the admission policy. The policy continued to be questioned by non-pool gins, bitterly now in light of Tahoka Cooperative Gin's recent decision to cease operations. Tahoka Coop, the non-pool gin in Tahoka, was going out of business. ACG was being held partially to blame. Boggs aired a request from Tahoka Farmers, the pool gin in Tahoka, that their neighbors from the Tahoka Coop Gin be allowed additional time to sign in or out of the pool due to the bankruptcy proceedings at their gin. The board agreed to allow them until May 31, 1985, to sign out of the pool or transfer to another pool gin.

Some of the non-pool gins were taking an increasingly radical view. They were beginning to object to pool gins being represented on the PCCA Board. Never mind that ACG members were members of PCCA. The attitude seemed to be, "If we can't participate in their pool, why should they be allowed to sit on our board?" Momentum appeared to be building toward two completely distinct cooperatives, one for pool gins and one for non-pool gins. Disgruntlement rapidly was turning to disdain. The board then joined the PCCA Executive Committee to discuss the policy.

Unfred opened the meeting, saying it was "regrettable" that problems had arisen, but he was "pleased the two groups were meeting to discuss it." Again cast in the role of peacemaker, he said, "whatever the resolution to the question, we all should continue to work in harmony, since PCCA, TELCOT, and ACG are now well established, and joint management of ACG and PCCA [is] economical for both cooperatives."

Boggs, in what was by now a tiresome exercise, reviewed the pool policy changes since the inception of ACG, and he made it a point to justify the pool gins being represented on the PCCA Board. He tried hard to find a way to explain the ACG Board's attitude toward expanding the pool. If he had to identify one key reason the ACG Board wanted to restrict the number of gins in the pool, he told the group, it was to limit the geographic area served because of increased membership communication required with a pool. He said there were gins in Oklahoma and South Texas that would like to have cotton in the pool, and it would be difficult to limit the area once cotton was accepted from outside the pool gins. That may have been true, but it was not a "key reason." Had he elaborated on the key reasons, it might have ended the meeting abruptly, or at least brought some tempers to a boil.

If the joint meeting was viewed as a public relations gesture, the effect was only temporary. Immediately after, the ACG Board cast its policy in concrete, unanimously adopting a resolution that said: "After meeting with the Executive Committee of the Board of Directors of PCCA and thoroughly discussing our policy of accepting pool cotton from the 27 pool gins only, we hereby reaffirm and continue that policy." To a man, the board held fast to its position. There would be no action taken on either side for another year, but the issue would fester just beneath the surface of relations between the two cooperatives.

ACG had become a participant in Levi Strauss & Co.'s "Just in Time" (JIT) program. The JIT Program was aimed at improving the system through which LS&CO drew on suppliers' inventories and shipments were made to their cutting plants. The goal was to keep the LS&CO's inventories, and the costs in carrying them, as low as possible. At the same time, it would provide LS&CO maximum flexibility in placing and receiving orders. LS&CO was trying to smooth the ups-and-downs of the denim market, for itself and its suppliers. In October, Bob Hale met in Wichita Falls, Texas, and Littlefield with LS&CO officials to talk about applying the JIT concept to ACG denim shipments going to LS&CO's Wichita Falls plant. By the end of 1985, with denim business good again, ACG's participation in the program was running as smoothly as the looms in Littlefield. And in December, meetings were to be held with LS&CO officials in Sedalia, Missouri, to set up a JIT program with that plant.

On December 11, the board met by telephone to discuss two members who had failed to deliver their cotton to the ACG pool. The cases show the close, personal nature the directors' role often assumed in dealing with fellow farmers. They also are interesting for their results. Mr. H. of Associated Cotton Growers delivered about 35 bales of cotton to an independent gin and owed Associated Cotton Growers $5500. Boggs had discussed the problem with T. W. Stockton and Mondell Mills, and they recommended that ACG file suit against H., requesting that he be forced to deliver the cotton to the pool. The second case involved Mr. D. of Midkiff Farmers Coop Gin. He had delivered cotton to Midland Coop Gin due to the ginning backlog at Midkiff. He was unaware, he said, of the requirement to deliver his cotton to Midkiff. Boggs talked with Midkiff Gin manager, Rocky King, who claimed that D. was aware that he was required to gin the cotton at a pool gin. The board moved to file suit against H. and postpone any action against D. until its December 19 meeting. At that meeting, Boggs reported that ACG had filed suit against H. for failure to deliver his cotton to the pool through Associated Cotton Growers and that a hearing was scheduled for the following day. He reviewed the D. situation, explaining that the cotton in question totaled 124 bales from one of four farms signed into the pool by D. Following discussion, a motion carried unanimously to refuse the 124 bales and cancel the contract with D., but only for the one farm in question.

The matter dragged into the following year. D. attended the January 16 board meeting and explained that he had made an "honest mistake" when he delivered his pool cotton to the non-pool Midland Gin. He claimed that of the 500 to 600 bales of cotton produced on that farm, only 124 had been delivered to the non-pool gin and promised to deliver the remaining cotton to the Midkiff Coop Gin. After hearing his statement, the board thanked and dismissed him. After he left the room, Dan Martin corroborated his statement, saying he already had ginned 384 bales from that one farm at Midkiff Coop Gin and that another 60 were waiting to be ginned there. Having heard the farmer's side and Martin's expression of support for him, the board voted unanimously to accept into the pool all of the bales that had been ginned or were to be ginned at Midkiff, but to refuse acceptance of the 124 bales ginned at Midland. The board recanted its earlier position to terminate the contract on that farm. All four of the farms that D. had signed into the pool would remain in the pool.

In two situations, the directors had taken firm action. In one case, they demanded that the farmer deliver his cotton to the pool gin. In the other, they refused to accept the farmer's cotton at the pool gin. No matter how detailed and formalized the procedures, decisions in a sensitive and highly personal environment like ACG tended to be partly intellectual, partly visceral, and totally complicated.

A four-month period of relative quiet at the beginning of 1986 only meant that pressure was building in the field and in the boards and committees. After a discussion of intensifying competition between pool and non-pool gins, the ACG Board again affirmed its policy of not allowing new gins into ACG. C. L. Boggs expressed deep concern over foment taking place in the field. Seven non-pool gins requested membership, two were considering merging with pool gins, and two others had gone out of business.

The 1986 membership meeting on April 24 was unremarkable. The terms of directors in Districts I and II were expiring, and the nominating caucuses submitted the names of the incumbents, Doyce Middlebrook and Jim Bob Curry. There were no nominations from the floor, and the two were reelected by acclamation. L. C. Unfred noted two milestones during the previous year, payment of the final installment on the original building bonds and attainment of the five-year financial goal. ACG would retire, for the third consecutive year, capital credits held by those members who reached their capital requirement.

On May 6, Unfred opened the board meeting by reviewing discussions at the textile pool committee meeting held earlier that day. It was the consensus of the committee to pursue a plan to limit the size of the ACG pool. The ACG Board began to be caught in a squeeze. On one side were the non-pool gins chafing at not being allowed to gin pool cotton; on the other its own textile pool committee was even more determined than the board to limit the pool. It was another blow aimed at the non-pool gins.

BATTLE LINES ARE DRAWN

The pool versus non-pool issue began to come to a head. Unfred told the board in a May 17 conference call that the non-pool gins had formed a committee to study the matter, and the committee had asked to meet with the PCCA Executive Committee and ACG Board jointly on Wednesday, May 21. The minutes of the conference call report Unfred said the ACG Board "probably needed to meet with these people and listen to what they had to say." In the understated way in which the board members expressed their views, Unfred was telling the other directors they had better have a meeting.

C. L. Boggs told the board what he had been able to find out. The group of non-pool gins had met in Slaton on May 8, with 30 to 35 gins represented. Charles Macha had been elected spokesman for the committee selected at the meeting. The group met again on May 15 and outlined their bill of particulars against ACG. (1) They were adamant in their belief that non-pool gins should be allowed to gin pool cotton. (2) They questioned whether a pool committeeman with all of his cotton in the pool should serve on the PCCA Board. (3) Aiming squarely at L. C. Unfred, they questioned the current joint chairmanship arrangement of PCCA and ACG. (4) And aiming at C. L. Boggs and his staff, they questioned the joint management arrangement between the two cooperatives. Jackie Mull, later PCCA Chairman, remembers the meeting in Slaton. "They were all open [non-pool] coop people, PCCA people, and when they met they made it clear that they were not interested in negotiation with ACG. They said, 'This is a PCCA deal.'" They wanted action, not talk.

Mull, who believes the "100 percent concept" was a mistake from the beginning, says that the problem "wasn't so much the ACG farmer against the PCCA farmer." In his view, "there wasn't any jealousy between them . . . because the PCCA fellow could go to a pool gin and join." The economic situation was poor, and "It was one gin against another . . . fighting over customers." Mull was one who argued that they should meet with ACG and talk about the problems, and they finally agreed. By this time, the ACG directors had come firmly to the opinion that they should meet with the committee from the non-pool gins and with the PCCA Executive Committee. The meeting was held on the morning of May 21, 1986.

Charles Macha spoke for the non-pool gins. He briefly reviewed the history of ACG, then presented the case of the non-pool gins.

He explained they were suffering from competition with the pool gins, and they had organized this committee to discuss and air their grievances and seek redress. What did they want? They wanted to be permitted to gin pool cotton. Other than their desire to survive economically, why did they think they should be permitted to gin pool cotton? In recognition of the financial assistance they gave ACG during its founding, through the regional cotton cooperatives. Macha then presented a petition to ACG. The petition, dated May 15, 1986, read as follows: "We the undersigned appeal to the proper authorities, who set ACG Board policy, to grant relief in a measure of fairness and cooperative philosophy to permit ACG members to gin at any coop gin that is an agent for PCCA." It was signed by 62 representatives of 44 gins. Unfred said the petition would have to be analyzed.

The next day, the monthly meeting of the ACG Board was held. Unfred distributed copies of the petition and noted that the non-pool gins had gained considerable support. Van May added that the entire PCCA Board, 160 strong, had adopted the petition in the form of a resolution, asking it be presented to the ACG Board. The stage then was set for a full confrontation. Unfred reported that the ACG Textile Pool Committee, the representatives of all the pool gins, had adopted a resolution running directly counter to the petition of the non-pool gins. By a vote of 19 to 1, the pool committee stated its opposition to allowing non-pool gins to gin pool cotton. Disgruntlement had become disdain. Disdain was becoming dissolution. Both sides having expressed their will, the battle lines clearly were drawn.

For 13 days, there was an uneasy quiet. Then, on June 3, Unfred called a special meeting of the ACG Board. He had some things to say. His ability to act as a peacemaker, as he had done so many times in the past, was being compromised. Not only was he caught in the middle of a situation that was becoming rancorous, half of his constituency now was questioning whether he should continue at all as the head of the two organizations.

Jack Hughes was there. He told the board he was convinced, after meeting with the non-pool gin group the day before, that there would have to be some change in the way PCCA and ACG operated. He said all efforts should be moving toward a single goal, that being to serve farmers and members of the cooperatives in the best and most cost-effective manner. Hughes conveyed many of the concerns he had heard from the non-pool group, emphasizing that their dissatisfaction had boiled over and now included unhappiness with the leadership of the two cooperatives. They had discussed, he said, the possibility of having separate board chairmen, of changing the PCCA Board so that ACG members could not belong to it, and of changing the joint management arrangement. No doubt about it. They were serious. Hughes said a compromise would have to be reached. He suggested a small committee be set up including representatives from both sides of the issue, from each regional cooperative, and with a coordinator who could be independent. For that role, he offered Ed Breihan, retired general manager of the Plains Cooperative Oil Mill. Breihan had been hired by the PCOM in 1975 to succeed John F. Herzer, who had left because of a difference of opinion over the contract sunflower program. Breihan was described in an October 9, 1975, article in the *Avalanche-Journal* as "widely known in the cotton industry and now serving as vice-president of the National Cotton Council." He would prove to be very helpful during the tumultuous time ahead for ACG.

Hughes also suggested an outside consulting firm might have to be engaged to look at the issues. He had a great deal at stake, personally and professionally, in seeing the issues resolved. The people in these factions were his long-time friends. They were among the finest cooperative farmers in the country, let alone Texas. And his bank had no small investment in their work and interest in their success. He implored the matter not be delayed and that positive action be taken as soon as possible.

Rex McKinney, the manager of the Farmer's Cooperative Compress, and Wayne Martin, manager of the Plains Cooperative Oil Mill, attended the meeting. They were asked for comments. Both cautioned that the problem, if not solved, could have a severe effect on all the regional cooperatives.

DIGGING IN THEIR HEELS

The board recessed briefly. The pressure was on them. They had worked hard, taken risks, pledged their assets, and succeeded. Now those who had watched from the sidelines wanted a piece of the trophy. The directors recalled the wishes of their textile pool committee. They plumbed their own convictions. And they made their decision. They would resist. They would circle the wagons and defend against the attack. The board reconvened. They said if they were forced to make a decision that day, the answer to the request presented by the PCCA Board would be that non-pool gins could not gin pool cotton. The cooperatives always seemed willing to talk in a dispute. The board said it would participate in a committee to review alternatives, as suggested by Hughes. Unfred asked Hughes to present the board's position to a joint meeting with the PCCA Executive Committee and the special committee from the non-pool gins to be held shortly after noon that same day.

The first order of business at the afternoon meeting was a correction to the minutes of the previous joint meeting. It came from the floor and was to clarify that the "non-pool gins were not concerned about competition with pool gins but about the ACG policies." The "correction" seems a transparent attempt at face-saving by someone in PCCA. In fact, the minutes of the ACG board meeting on March 21 indicate that Charles Macha did say the non-pool gins were suffering from competition with the pool gins. Could anyone really believe that competition was not the issue, that the non-pool gins simply were upset that ACG made the policy it did? Had ACG been financially troubled and the non-pool gins successful, would the non-pool gins even care what ACG's policy was? Probably not. Nonetheless, the correction was made.

Unfred called on Charles Macha. Once again, he stated the desire of the non-pool gins for pool members to be able to gin at the cooperative gin of their choice. He said if the problem could not be solved, the alternative might be splitting the two cooperatives.

Jack Hughes was called on to speak. He repeated his comments before the ACG Board earlier in the day, including his suggestion that a committee be formed to review the situation. Then he went into additional detail. He suggested the committee might include the chairman of each regional cooperative, three non-pool farmers, three pool farmers, and a moderator, who must be an outside third party. He tried to explain to the group that the ACG Board was not taking a firm stand on the pool only for itself. Rather, it did not feel free to change the ginning policy because the resolution passed by the textile pool committee insisted the policy not be changed.

One of the PCCA Executive Committee members put the question directly. "Why," he asked, "can't non-pool gins gin pool cotton?" An ACG Board member was vague in response. He said there were some "questions about representation that needed to be answered" and also the textile pool committee resolution "really tied the ACG Board's hands." The ACG Board never had been eager to expand the pool, other than when the bank made it imperative. Nor had it ever been quick to allow non-pool gins to gin pool cotton, other than when pool gins could not handle the load themselves. Now, trying to explain its actions, it fell back on the position of the textile pool committee, which, if anything, was even less willing to bend. There was lively discussion and some questions from the floor. Lashing out at one of the most visible symbols of the relationship, a non-pool committeeman suggested that ACG and PCCA needed separate board chairmen. Sensing what was coming, the suggestion of a complete split, Hughes reacted immediately. He told the group a total split of the two cooperatives might cost each an extra million dollars a year in expenses. There was more discussion. The beleaguered Unfred recounted the benefits of having a joint staff between PCCA and ACG.

The ACG Board recessed for 30 minutes and reconvened. Macha bluntly put the question to the directors again. He asked for a definite answer not later than the June 18 PCCA board meeting to the resolution put before them by the PCCA Board. He said the non-pool gins would wait until that meeting to take further action. That gave ACG 15 days to respond. Unfred suggested it would help if Hughes's committee suggestion were put into action before June 18. He was concerned that to let the situation slide until then might result in the PCCA Board taking harsh action from which it might be impossible to recover. Hughes said perhaps one pool and one non-pool farmer from each regional could be selected immediately for a committee. Then he went as far as anyone could expect the president of the Texas Bank for Cooperatives to go in helping the groups reach a solution. He suggested that Wayne Martin, Rex McKinney, and he could select the committee representatives, if that was the desire of everyone involved.

No further action was taken in joint meeting. June 3, 1986, had been a long day. Their nerves on edge and emotions drained, all the groups but one adjourned to gather their thoughts and plan their next moves. The remaining group was the ACG Board. It reconvened immediately after the joint meeting, long enough to decide to call the textile pool committee into session at 9:30 a.m., June 18, early on the day of the PCCA board meeting. The directors wanted the committee's views once again before making a final decision on the PCCA Board's request. They also wanted to ensure that the members of the committee fully understood the gravity of the situation.

On Friday, June 13, C. L. Boggs convened the board. He said that two days earlier some non-pool gin representatives had presented a letter to the PCCA staff. They asked it be mailed to all PCCA directors and gin managers. The letter took a hard line against the ACG position and threatened separation of the cooperatives. The PCCA Executive Committee, Boggs said, stopped the action. They directed the staff not to mail the letter, at least not then. They still were waiting to hear the ACG Board's final decision. There were some cool heads among the non-pool gin representatives, and they did not want PCCA torn apart.

On the morning of the 18th, L. C. Unfred told the ACG Board and members of the pool committee that, if the issue led to dissolution of the joint management agreement, annual operating costs for each cooperative would increase by nearly one million dollars. The agreement, he said, had been fair and equitable to both sides. If it would help resolve the issue, he said, he and Doyce Middlebrook had agreed they would ask not to be considered for reelection to the positions of chairman and vice-chairman when the PCCA Board was chosen in September. Having said everything he could, he then called on Jack Hughes.

Hughes emphasized he was not attending as president of the Texas Bank for Cooperatives, but rather "as an individual" to clarify what could happen if the issue were not resolved. He said he tried to be candid during meetings with pool and non-pool gins, telling them that separation of the regional cooperatives *must* be prevented. He had stressed to the non-pool committee that rationality *must* prevail during the PCCA board meetings that afternoon and urged them to consider all possible compromises to settle the issue. He reiterated, this time for the textile pool committee, that it would be in the best interest of both organizations to form a joint committee to work together in order to find a solution.

Following comments from Rex McKinney and Wayne Martin, the textile pool committee moved ever so slightly. A motion carried to recommend to the ACG Board that "policy requiring pool cotton to be ginned at pool gins not be changed for the 1986 crop but that a committee be selected to meet with PCCA representatives to try and resolve the issues." Even on this, there was one dissenting vote.

Immediately after the meeting with the textile pool committee, the board convened and unanimously adopted the following resolution:

> RESOLVED, that American Cotton Growers does not wish to sever its relationships with Plains Cotton Cooperative Association because American Cotton Growers believes that working together in harmony with Plains

> Cotton Cooperative Association over the past eleven years has greatly benefited the cotton farmers of both organizations, and that American Cotton Growers would like to participate in discussions about the future of these two producer cooperatives through a duly appointed committee made up of representatives from both organizations.
>
> Be it further resolved that American Cotton Growers is unable to change its policy requiring that the 1986 crop pool cotton be ginned only at designated pool gin locations, but that American Cotton Growers is willing to discuss all future alternatives within a committee framework in order that ACG and PCCA may continue to work together for the best interest of cotton farmers served by these organizations.

The Hughes approach was about to be adopted.

THE CONFLICT GOES TO COMMITTEE

There was strong feeling all along that Hughes's approach would be adopted. The cooperatives tended to listen carefully to the bank, and Hughes was respected. In the days since June 3, when L. C. Unfred suggested that work begin immediately on developing the structure for the Special Study Committee, the groups selected their representatives. The representatives from PCCA were: District I — Kenneth Burnett of Cotton Center; District II — Tex Martin of Kress-Swisher; District III — Charles Macha of Opdyke; District IV — Jackie Mull of Idalou; District V — Joe Rankin of Owens; Rolling Plains Cooperative Compress (RPCC) — Douglas Church of Stanton; Oklahoma Cotton Cooperative Association (OCCA) — Coy Grimes of Humphreys; Rio Grande Valley (RGV) — Harold Scaief of San Benito; Coastal Bend — Curtis Jensen of Danevang. From ACG: District I — Raymond Belew of South Smyer; District II — Larry Lockwood of Spade; District III — LeeRoy McCravey of Yoakum County; District IV—Wayne Huffaker of Tahoka; District V — L. C. Unfred of New Home; District VI — Wendell Jones of Glasscock County; District VII — Compton Cornelius of Associated Cotton Growers. The name Unfred leaps from the page. The chairman of PCCA, the great majority of whose members were non-pool gins, would represent his home district on the side of the pool gins of American Cotton Growers. Just as T. W. Stockton had cast his lot with Crosbyton years earlier, Unfred now would ally himself with those closest to home.

The ACG Board went into the PCCA Board meeting prepared to listen and to deal. Charles Macha and Compton Cornelius were elected spokesmen for their respective sides. One of the committee's first actions was to hire the consulting firm of Emmer and Associates to study the issues. Based in Evergreen, Colorado, Emmer was recommended as a firm that had worked with agricultural cooperatives in the past.

By mid-July, the committee had chosen Ed Breihan as moderator and held two meetings. The ACG representatives offered to recommend to their board that growers who signed new farms into the pool after April, 1986, be permitted to gin these farms at the cooperative gin of their choice. Having made the offer, they then withdrew it pending completion of the Emmer study. Emmer was to begin the study on July 22, and estimated that a minimum of three to four weeks would be needed to complete it. The ACG Board pondered its role with regard to the study. The directors decided they would pay one-half of the reasonable cost of the consultant's study; they would monitor progress and cost; they would set guidelines for ACG representatives on the Special Study Committee, if needed; and any recommendations made by the consultant would have to be approved by them. The study committee designated six areas for review by the consultants: (1) the allocation of costs between the cooperatives, (2) the allocation of sales between the cooperatives, (3) whether the staff solicited members for ACG, (4) review of the board structures and policy formulation procedures of the cooperatives, (5) whether management had a conflict of interest, (6) whether there was a conflict of interest between the two organizations.

By mid-August, Unfred told the board the consultants had found nothing wrong with the joint management of the two associations. The consultants recommended, however, that the members of each cooperative look at the potential of merging some of the cooperatives in West Texas. In this regard, they would have found a supporter in Jack Hughes.

On August 21, Ed Breihan and consultants Gerald Emmer and Al Lambrecht attended the monthly ACG board meeting. Jim Bob Curry asked if, in their opinion, there was a problem with the joint chairmanship of PCCA and ACG. Breihan said the question really had not been addressed by the committee. They had not focused on Unfred's role. Harold Barrett asked if there were any question about the ability of ACG members to serve on both boards. Breihan responded that there had been quite a bit of discussion about that, but the discussion centered around key points of the marketing agreement between PCCA and ACG.

ACG Versus PCCA

UNFRED PAYS THE PRICE

If anyone suspected it, they did not say so, but the curtain surely was lowering on L. C. Unfred's chairmanship of PCCA. Like his predecessor, Howard Alford, who had served as president of PCCA and Farmers Cooperative Compress, Unfred had led two regional cooperatives. He led them through a dynamic era, and both owed much of their success to his patience and skill. Ironically, the two organizations he served with such diplomacy would find it virtually impossible to declare a truce with each other. It was only a matter of time before the impact would fall on his shoulders.

"I think L. C. understood the non-pool gins' problems," says Dan Martin, "and he was more concerned about the well-being of non-pool gins than most people realized. I credit L. C. with keeping the coops together. He was the glue. Had Howard Alford still been here, PCCA and ACG would have split right down the middle, because he was so outspoken."

Whether Martin's conjecture is accurate, Unfred's sensitivity to the political nature of the cooperatives gave him a hint of the rejection he would experience in the 1986 PCCA annual election on September 17. Early in the summer, though no one had asked them to, both he and Doyce Middlebrook had offered to step down from their leadership posts in PCCA. The fact that no one had asked for Unfred's resignation, nor even suggested he resign, was symbolic of the respect in which he was held by the members. If he were going to be removed from office in PCCA, it would be done fairly, through a vote of the board. And so it was. Of 160 PCCA board members, five more voted against Unfred than for him. The vote was not a resounding defeat, nor was it a sign of dissatisfaction with Unfred's leadership. Instead, it was a clear signal sent by PCCA to her sister cooperative, a strong expression of opinion regarding ACG policy.

Unfred passed the gavel to Jackie Mull, a producer from the Idalou Cooperative Gin, a non-pool gin, who sat on the Special Study Committee. Mull is a scholarly looking man with bulldog determination. He had nearly died years earlier, the result of a heart attack. He had been overweight and a chain smoker. With strong self-discipline and rigorous exercise and diet, he now looks 10 years younger than his 51 years. He had run against Unfred twice before, but not because of any disagreement with him or dislike for him. "I think L. C. did a good job," he says. "I had no problems with him." Mull opposed him because he was asked to by some non-pool members, and because he felt competition for the office was healthy for the organization. As for those non-pool members who encouraged him to run, he says, "They felt because [L. C.], as [an] ACG [member] and chairman of the PCCA Board, he was in a position that he would have a little more input into ACG. He was in a position to know what ACG was doing and what PCCA was doing. The way [ACG's] board was set up, there was not much information let out. The reasons they made certain decisions were kept within that board. I'm not saying that's wrong," he adds. "It's just the way it was."

Mull knows that now from experience. PCCA's 160-man board was later reduced to 11, largely because of security concerns. Mull says, "When you are in the denim business, there is information you don't want out in public. If you tell 160 men, it's going to get outside." He says the relationship with Levi Strauss & Co. is a major factor. Trade secrets have to be kept, or strong business relationships cannot be maintained.

Despite his having been several times the candidate of the non-pool members, Mull was not slavish to them. To the contrary, he had taken the position after the Slaton meeting in May that the gins there "were not speaking for PCCA, that they were speaking for themselves," and that they had turned the issue into a PCCA–ACG disagreement. He was a strong supporter of PCCA, but "took the attitude that ACG was a coop organization of its own and . . . I had no business telling them how to run their business." He also knew that if PCCA and ACG split completely, "we would have lost a lot of money that they were paying us for management. It would have just about ruined PCCA."

He acknowledges that Unfred was in an impossible position. "We were going through these study committees and arguments between the two companies . . . and they were plenty hostile arguments at times. We'd sit down at a table, and one side would be ACG and the other side would be PCCA. We would argue back and forth about what we [PCCA] thought they ought to do to make things more equal and fair. L. C. was chairing those committees and, of course, he would side with the ACG group . . . I can't say anything is wrong with that because he had to take a position on one side or the other." Hard as he tried, even as skilled a diplomat as Unfred could not appear completely objective to both sides. The more contentious the situation became, the hotter it was for him in the middle.

Even with the level of dissatisfaction as it was, Mull did not expect to win the election. "In fact, whenever they asked me if I'd run for chairman I didn't campaign against L. C. I just ran. And I was floored when I was elected, because I just knew he would be reelected." In the earlier contests, Unfred had enough support from non-pool ginners to win handily. By 1986, that was no longer possible, and he lost by the smallest of margins.

As Unfred left his post, the PCCA Board passed a resolution that is testimony to his 13 years of labor on behalf of its members. The resolution cites his service to "fellow cotton producers throughout Texas and Oklahoma with untiring diligence and distinction . . . [and his devotion of] considerable time to provide leadership and further the interests of his fellow cotton producers." It recounted his leadership in development and implementation of the TELCOT system that "enabled producers to attain a competitive edge in marketing their cotton," and his "unselfish" representation of his fellow producers on the boards of numerous national cotton organizations. Calling him "respected and widely admired" in his community, Texas, and Oklahoma ". . . an agriculturalist, a cooperator and a leader," the PCCA Board extended him its "sincerest gratitude and deepest appreciation."

THE COMMITTEE REPORTS

At the August 21 board meeting, Ed Breihan delivered the Special Study Committee report. He thanked the committee for asking him to serve as its moderator. He said it was an "interesting" experience. That is a word often used by persons who are asked to describe something that is particularly difficult or unpleasant but about which they feel compelled to be gracious. He recounted the record of accomplishment achieved by the two groups since 1975:

1. PCCA net worth has grown from $13 million then to $28 million today. Working capital has grown from $4.5 million to $14 million.
2. The PCCA TELCOT system has been built into a very important marketing innovation for farmers.
3. ACG net worth has been increased from $3 million in 1975 to $38 million today. The $6 million which was invested in cash by member gins and three Lubbock regionals has been repaid.
4. The denim mill is working fine and its output is being marketed in a very profitable manner.
5. Through the years, the cooperatives working together have developed a substantial amount of political "clout." This has been important and will continue to be important as you move forward. All of this speaks very

> well of the cooperatives themselves and the joint management structure. We should think a long time before we break up something that has worked so well.

He credited the committee with attacking the issues with diligence and said the members had met in some form every week over the preceding six weeks since they were appointed. He introduced the principals of Emmer and Associates, emphasizing the importance of an independent, objective view, which the use of outside consultants added to the committee's work. In addition to the six areas the committee and consultants were to study, the consultants were directed to pursue any items related to the issues and to report to the committee. The committee had asked the PCCA and ACG staff for reports in two areas. One was the 1986 ACG sign-in-sign-out acreage, to determine more accurately the location and source of new acres signing into the pool. The other was the estimated effect on the expenses of the cooperatives in the event of a complete break. The auditors of both cooperatives and the consultants checked the numbers provided by the staff.

Breihan reported the findings and recommendations of the study. The conclusion on allocation of sales and costs was that it was being done in a fair and professional manner. The committee recommended no significant changes. The current marketing agreement between PCCA and ACG had been reviewed. It was found simple, straightforward, and generally fair and effective over the years. It addressed the PCCA membership of all ACG members; the requirement that all ACG cotton be marketed through PCCA; and the basis for the cost sharing arrangement used by the organizations. If changes were considered, the committee cautioned, the entire agreement might need to be renegotiated to maintain balance.

On the question of whether staff solicited members for ACG, the consultants interviewed a number of people. They reminded the Special Study Committee that, not too many years earlier, it was the unwritten policy of PCCA to encourage membership in ACG. The current controversy, they found, was putting staff members in a very delicate position. But they found no substantive evidence of staff favoritism of either cooperative. In fact, they felt the staff went out of its way to avoid favoritism or even the appearance of favoritism.

The issue of management conflict of interest was addressed in substantial detail. The consultants reported to the committee that management had a conflict of interest, by definition, if the cooperatives viewed their best interests as being in conflict. The relevant question, they said, is, "how is such conflict handled?" The consultants had several suggestions to help the boards ensure no management favoritism developed in the future.

1. Each cooperative should form a small audit committee to determine the scope of its audit and receive the auditor's report.
2. Each cooperative should form a small compensation committee to handle the question of management compensation. The consultants felt the matter of determining how management was compensated was not being handled in a professional manner. "The present system," they said, "gives rise to unnecessary suspicions that management is being encouraged to favor one organization over the other." The Special Study Committee voted to recommend the ACG Board and the PCCA Executive Committee serve as compensation committees. They would have responsibility for studying comparability with similar organizations; evaluating management performance; and coordinating with each other. Each would set, rather than recommend, management compensation for its own cooperative.
3. PCCA should consider somehow reducing the number of voting members of its board. The consultants' experience, from working with many cooperatives over the years, was that sheer numbers made it impossible for the board adequately to understand, question, and evaluate management proposals. "Reducing the number of voting directors," they said, "to 25 or less from the present 160 could make the board more effective." The Special Study Committee voted unanimously to defer this question. They felt it could be considered later, along with other issues of structure.

Breihan reminded the ACG Board that, in many respects, the study became necessary because of the controversy over the ginning policy associated with new acres signed into the pool that year. "The consultants feel that there is conflict of interest between organizations all right," he said, "but basically it is not in PCCA or ACG as marketing organizations." He placed the conflict exactly where Jackie Mull had seen it. "The conflict," he said, "is at the gin level." As marketing and processing cooperatives, PCCA's success does not hurt ACG, and failure on the part of ACG does not help PCCA. Thus, one must conclude that they have no basic conflict of interest. "Using that same test," he added, "even affiliated individual producers have no basic conflict of interest. A PCCA producer does not benefit if an ACG producer gets less for his cotton and vice versa. Because of the direct correlation between gin volume and gin profitability, however, competing gins do have a very real conflict of interest. To the extent that ACG policy affects the ginning location of pool cotton, both ACG and PCCA are drawn into the conflict. This is exactly what has happened."

The consultants recommended actions to resolve the issue. First, the ACG Board should issue a statement of its intent regarding pool reopening. It should do this as soon as possible. The statement should say that the pool is now closed, with no intention of reopening, unless the size proves inadequate to optimize mill operations. It should state that it is ACG's intention that, if and when the pool is reopened, clear priorities to entry will be set and the 1984 restrictive ginning requirement will be addressed for nonaffiliated ACG acres entering the pool. It should express ACG's intention that future changes in the policy will be discussed with the PCCA Executive Committee for its participation and reaction before finalization.

Second, the ACG Board was encouraged to discuss the pool reopening issue and develop a comprehensive policy with the following considered:

1. A restatement of the primary goal of ACG as it was formulated on March 19, 1975. "One major objective of American Cotton Growers and of the cooperative gins that participate in the textile pool," it said, "is to develop a stable membership of cotton producers who share the same cooperative principles and goals."
2. Upon reopening of the pool, an acreage cap will be established that reflects the optimal pool size to assure a constant and adequate supply of cotton to keep the textile mill operating as economically as feasible.
3. Upon reopening, priority will be given to additional acreage from those farmers who were in the pool at last sign-up and now have new acres to include.
4. If the pool cap still has not been reached, the next priority will be farmers who are members of an ACG-affiliated gin, but did not have acreage in the 1986 sign-up.
5. All acreage signed under the first and second priority may be required to be ginned at an ACG-affiliated (pool) gin.
6. If the sign-up still is short of the optimal pool cap, it may be offered to other farmers. Acreage signed in under this offering would not be required to be ginned at an ACG-affiliated gin. Furthermore, such new acreage from nonaffiliated cooperative gins may be required to make an initial investment in ACG. This initial investment level would be set to reflect, in part (though it does not say to what extent) the time and cost of the original investment capital so as to bring some measure of equity to all invested parties.

The consultants had crafted a carefully phased plan to protect the interests of the pool gins and to bring the non-pool gins into the process at a specific point, that being failure to reach the optimal pool cap by (1) farmers who were in the pool at the last sign-up and now have new acres to include; and (2) farmers who gin at pool gins but did not have acreage in the last sign-up.

The Special Study Committee had voted unanimously to accept the consultants' recommendations and present them to the ACG Board. The study included a section on future issues that should be considered by the cooperatives. "This study of . . . six issues, as important as it is, has served to remind us of the need for strategic planning for the future. Cotton producers in Texas and Oklahoma need much more help from the cooperatives. The cooperative ginning industry is locked in a competitive battle for volume that promises to be a destructive force in the industry for years to come.

"The need for this study has reminded us that it probably is time to look at ways to improve the structure of our cooperatives. The Farm Credit System waited until it was in real trouble to make such a study. Now it is forced to consider change when much of its resources are gone. Let us not do the same." The committee went beyond the bounds of its assignment, but with reason. It hoped to avoid recurrences of the controversy it had been established to resolve. Its recommendations, only one of which was implemented and then not in the form envisioned by the committee, are included here because they are noteworthy for future consideration of the structure and operations of the Lubbock regional cooperatives.

"Our consultants," the report said, "have suggested some possible restructuring proposals that the committee feels are worth considering. It will take time but we might start thinking of such things as:

1. Consolidating PCCA and ACG, making the textile mill pool a division or a subsidiary with its earnings staying with its present members.
2. Consolidating the compresses with the new marketing cooperative. (The advantages of this would be that any [unencumbered] working capital could be refunded to farmers; it would eliminate much of the cost of duplicated operations such as electronic data processing; and it would use the earnings of the compress to level out the peaks and valleys of marketing earnings.)
3. Offer gins the opportunity to consolidate with the new organization.
   a. Assure managers of continued employment with the new organization.
   b. Keep gin point offices open.
   c. Consolidate actual ginning at strategic locations.
   d. Move quickly toward the most efficient seed-cotton transportation systems.
   e. Allow gins that want to remain independent cooperatives to do so. If they want to consolidate to join the program and then reorganize as

> new cooperatives after modernizing their locations and operations, they could be allowed to do so.

The important thing for farmers is a modern and efficient system and the dollars such a system can return to them.

The oil mill, the study said, may be a different enough business that it need not be a direct part of the new consolidated operation. Waxing philosophical, the committee added, "But who knows what may be advisable further down the road?" They recommended the PCCA and ACG boards give them or another committee "the authority to pursue the advantages and disadvantages of the greater structural considerations. . . . This new charge should be specific and a target date set for a final report back to both organizations."

The topic of restructuring, in addition to being a particular favorite of Jack Hughes, surfaced whenever discussion turned to the roles of the regionals. Wayne Huffaker says, "I've supported that for years. It's ridiculous that [now] we've got three separate organizations [the Lubbock regional coops – PCCA, PCOM, and FCC] working for the same people." He has a theory about why this is the case and why it is not likely to change. "Everybody is protecting his turf . . . management, boards of directors, presidents. . . . If [they are] merged, everybody's figured out that there's just going to be one board and one president and there'll be drastic changes. So I don't know whether they'll ever get [restructuring] or not. One of the things that really burned me," Huffaker says, "was that [PCCA] has got a $30 million computer system, or however much the system cost, and the compress needed a new computer system, so they went out and spent $4 million or $5 million putting in a computer system over there when we've got a huge one here, state-of-the-art. That burns my ass bad." There had been little interest in 1980 when Hughes talked of the benefits of the regionals combining to "group their assets." There did not appear to be any more interest in it in 1986. Even if there were, it soon would have been overshadowed by actions involving ACG and PCCA.

The Emmer study prescribed that "PCCA and ACG keep moving forward in basically the present pattern, resolve the current conflicts as amicably as possible, and try to continue the successful operations of both cooperatives." The Special Study Committee and their consultants had made a good try. But the "present pattern" was completely unacceptable to the non-pool gins.

In a telephone meeting of the ACG Board on September 3, C. L. Boggs said the staff had heard within the last week that some gins might file a lawsuit in an attempt to force ACG to change its policy on pool membership. Very little was known about what the suit would entail and there was little the association's attorney could do without more information. Boggs said many gins had been contacted by the staff, but the strong feelings seemed limited to only a few gins.

POOL POLICY, AD NAUSEUM

ACG decided to take advantage of Emmer and Associates' experience, including their newly acquired knowledge of the West Texas cotton cooperatives, to help develop a plan for limiting the size of the pool. On September 9, the board voted unanimously to hire them, and on October 1, Gerald Emmer and Al Lambrecht attended a board meeting. The target for completion of the latest study was the end of October. The board felt the pool was large enough at the time, perhaps too large, and there was no need necessarily to rebuild it to the current level when members dropped out. A key issue the directors agreed on was the necessity for existing pool members to be able to expand their farming operations and place additional acres into the pool in the future.

By October 30, Emmer and Associates presented a preliminary report. Jerry Emmer said several issues needed discussion. He and Lambrecht tried to find precedents. They communicated with other organizations maintaining pool marketing operations. They found ACG's pool was unique, and a policy would have to be tailored to fit ACG's specific situation. He presented copies of the preliminary report to the board. Emmer's recommendations involved the transfer of pool membership rights, a requirement for advance capital investment on new pool bales, the allocation to gins of pool acreage for ginning, and the requirement that pool cotton be ginned at pool gins. The board decided to present the entire report to the textile pool committee prior to endorsing it. On the following day, the textile pool committee met to look at the report. Their attention went immediately to a recommendation that non-pool gins be allowed to gin cotton from additional farms signed into the program. They didn't like it. The board removed it. L. C. Unfred recognized that a lot of work remained to be done on the pool policy. The board instructed Emmer and Associates to continue working, and they did. The October 31 deadline Emmer originally set for completion of the study was long past. They worked through the end of the year and into the beginning of 1987.

Like perennials, discussions of the pool membership policy bloomed every year; 1987 was no different. At a February 16, 1987, board meeting, Emmer and Lambrecht reviewed their proposal, which had been revised once again. Emmer encouraged the board to review it periodically and change it if necessary. Urging the ACG Board to review the pool membership policy was like urging a fish to swim.

The next afternoon, Lambrecht presented the policy in a joint meeting of the ACG Board and the PCCA Executive Committee. The pool would remain closed to new members, "except in certain circumstances." The ACG Board reserved the right to grant membership rights to new members and replace bales lost to the pool through discontinued production by current members. But it clearly stated that this "is not contemplated until such time as the pool falls below its target size." The board could grant membership rights to new landlords or tenants on farms where they joined landlords or tenants who were members as of April 30, 1986. They also could grant membership rights to farms purchased from members who were in the pool as of that date, as long as the election was made to keep the farm in the pool by the sign-up date following the purchase. Such a farm would be able to remain in the pool even if resold later. And the board was empowered to transfer membership rights from members who wanted to cut back or discontinue production to their immediate family members. The farms then in the pool could stay in it indefinitely, regardless of who might own or farm them in the future. The pool was closed to new members except in cases where a non-pool grower affiliated with a pool member by virtue of a change in rental arrangement or land ownership.

The ACG Board, the proposal said, "shall establish a target pool bale count of 350,000 bales which reflects the current optimal pool size assuring a constant and adequate supply of cotton to keep the textile mill and ACG's cotton marketing efforts operating as economically as possible." Those eligible for pool membership would be permitted to expand their bale count. And in case the point was not made forcefully enough, the proposal restated that, except for those eligible, "the ACG Board shall *not* reopen the pool to new members until such time as the actual pool bale count declines to less than the target bale count through attrition." The authority to transfer rights to the pool from one member to another would rest exclusively with the ACG Board. "Current members of ACG would be able to expand or change their farming operation however they saw fit and bring any new production into the pool if they so desire.'' When the board's guidelines conflicted with one another, it reserved the right to use its ''sole discretion'' in deciding which guideline would take precedence. It would ''consider such factors as a proposed member's cooperative philosophy and goals, his production location, cost of production differences among areas, the needs of individual gins, and other factors in selecting members. . . .'' The ACG Board had absolute power. Membership would be truly exclusive. In the event the pool ever was opened to new acreage, the board would, ''give first consideration to producers located within the pool gin communities which lost acreage. [If] adequate, suitable acreage cannot be signed in from those . . . second priority shall be given to producers located within other pool gin communities, [and] in the event insufficient acreage is obtained [from those], the board may offer pool membership rights to other interested farmers located outside of an existing pool gin community.''

All these strictures really were small bones of contention. PCCA was ambiguous about ACG wanting to keep its pool closed. On one hand, many members of PCCA were disappointed when they tried to join the pool and were rebuffed. On the other hand, the non-pool gins were pleased their members could not just leave them and go to pool gins. The *big* problem continued to be ACG's insistence on not allowing its members to gin at non-pool gins. The proposal addressed that issue next. ''All pool cotton currently being ginned at pool gins,'' it said, ''and all new member cotton, as well as any new cotton entering the pool from current members, shall be ginned at pool gins. In the event a pool gin goes out of business, its pool members who desire to stay in the pool must transfer their ACG membership to another pool gin.'' Not only were members required to gin at pool gins, but if their pool gin folded, they would have to travel to another pool gin or leave the pool. The policy was not calculated to warm the hearts of the non-pool gins, and there was nothing ambiguous about it.

The final point of the proposal seemed to recognize that the joint management arrangement might not survive much longer. ''[This pool] policy,'' it said, ''was developed in consideration of the continuance of the joint management and shared marketing arrangement . . . employed by PCCA and ACG. Should the . . . arrangement be altered or eliminated, immediate changes in this policy may become necessary so as to keep [it] responsive to the best interests of ACG, its farmer members, and their affiliated gins.''

Following Lambrecht's presentation, PCCA Chairman Jackie Mull called for his executive committee to caucus. When they emerged, Mull said it was their opinion the policy was improved over the previous one. That was because transfers from non-pool gins to pool gins in order to get into the pool had been greatly restricted. But, he said, they were "disappointed" that the primary issue, that of not allowing pool cotton to be ginned at non-pool gins, had not been changed. With the graciousness characteristic of the people involved, Mull said he "surely hoped that PCCA and ACG could continue to work together in harmony and to continue their joint management arrangement."

Unfred was equally courtly on behalf of ACG. He reiterated the hope that the two organizations could continue to work together. A hint of frustration in his voice, he said the ACG Board and pool committee had spent many hours discussing the issue and believed "the policy as adopted seems appropriate for ACG." Both organizations had traveled very far to make very little progress.

The staff already had mailed copies of the pool membership policy to members of the ACG Board, textile pool committee, and PCCA Executive Committee. Several non-pool members and gin managers called Boggs and his staff with questions. He asked the board's permission to distribute copies to non-pool gin managers and PCCA board members, and it was granted.

For three years at this point, the controversy between the pool and non-pool gins had festered like an open wound. Boggs was not optimistic that the situation would get better. "It was very frustrating," he says, "and I was about at the point of believing that there was no solution to the problem."

Climax

LOCOWEED

The afternoon and early evening of February 3, 1987, was an agonizing time for Van May. He and Boggs were scheduled to fly to Denver the next day for a meeting with Emmer and Associates. The topic was the final plan to close the ACG pool. May was consumed with worry about the implications of the measure. He knew slamming the door on the pool was the obvious solution, but it posed serious political questions. He was convinced it could be the straw that would break the cooperative camel's back. His concerns were compounded by five consecutive years of poor cotton crops on the Plains. Coupled with sagging cotton prices during much of the period, they had made shambles of the West Texas and Oklahoma agricultural economy. Bankruptcies among farmers and agribusinessmen in Lubbock were at an all-time high. More deeply disturbing was the number of farmers May knew personally who were going broke.

Alone, in the quiet of his office at PCCA, May reached for a copy of the ACG financial statement and began reviewing the document. Eventually, his eyes rested on the line showing the denim mill valued on the books at $30 million. "The balance sheet reflected the depreciated book value of the denim plant," he says. "I felt the market value was well in excess of that. If ACG were a publicly traded company and it was common knowledge that they had a plant worth millions of dollars more than it was on the books for, it would be reflected in the price of their stock. Members would have the ability in the open market to cash in on that off-balance sheet asset. But as a cooperative, it wasn't structured that way. . . . Further, as a cooperative, we had guys who had these big investments in ACG that they had accumulated during the period of the small pools and nice earnings, who were not being remunerated for what they had invested, because as a coop what we do every year stands alone, and whoever delivered cotton to the pool in 1986 was going to get their 1986 margins. All these guys were going to get ultimately was their stock retired. . . . I began to think, if I was a farmer struggling like so many of them were in 1986 and I could say 'What's best for me out of this cooperative?', what would I like to see happen? The answer was that I would like a way to tap my equity to reflect the real value of the assets, not the value on a piece of paper." Ideas took shape in his mind. He knew any businessman on the verge of going broke would sell any stock he owned in a corporation to bolster his cash flow. Selling the mill might be a logical business decision for ACG members who were in financial straits. May insists the idea was not instigated by the need to resolve the pool versus non-pool controversy. Instead, he says, it began as recognition of the mill and its value as a resource to assist financially strapped ACG members.

But the pool issue, and the pressure under which it put the staff, certainly started him thinking. Like Boggs, May had seen the issue arise time and time again. Like some mythological monster, each time it arose it was bigger, uglier, more frightening, more powerful. Like Boggs, May was concerned that, perhaps the next time, it would destroy the relationship between ACG and PCCA. The results would be catastrophic for the cotton cooperative family, adding increased operating costs at a time when members needed as much cash as they could get from their regional cooperatives. He held the idea in his mind, turning it around and around to examine each side. His confidence grew that he had found the evasive solution. Long after most employees had gone home that evening, he was pacing in his office or sitting behind his desk making notes. He did this for three hours — reviewing, writing, analyzing, thinking.

The next day, May and Boggs boarded an Aspen Airlines flight at Lubbock International Airport for the short flight to Denver and their meeting with Emmer and Associates. As the jet reached its cruising altitude, the two executives began to relax. It was then, in the quiet of the plane and with a captive C. L. Boggs for an audience, that May sprang his idea. "Van said to me, 'Maybe we ought to just consider selling this denim plant,'" Boggs recalls, "'It seems to be at the center of the controversy we're in . . . the denim market is extremely strong now, so it probably has the highest market value it's ever had. Most people sell a thing when it reaches a good point in time to sell, and maybe we ought to consider selling it.''' Boggs was stunned. "My first reaction was that Van must have been eating locoweed, because to sell something that had been as successful as that denim mill was just inconceivable." As the airplane began its descent into Stapleton Airport, Boggs and May agreed to say nothing about the subject to Emmer and Associates.

One of Boggs's characteristics is his ability to focus on the subject at hand and eliminate distractions. It was hard for him to do this. May's words kept replaying in his mind as the consultants presented their final recommendations on how ACG might close the pool. On the return flight, the two again discussed the idea of selling the mill. Boggs muttered something about the possibility of selling it to PCCA. This time, May thought Boggs was crazy. Textile manufacturers, not other farmers, were the likely buyers May had in mind. Unlikely as it was that PCCA either would be interested in buying the mill or could find financing to do so, Boggs mused about keeping it under cooperative ownership.

The two would share their idea of selling the denim mill with no one for the next few weeks, not even others on the staff. In fact, the staff would see little of Boggs or May during this period. They stayed in their offices to discuss the idea of selling the mill, what the effects might be, and how it might be done. They decided a good starting point would be to determine the value of the plant. Boggs says he had "no idea" what it was worth and adds, "I wasn't about to go to the board to mention this unless I had some idea what it was worth because they would have thought I was nuts, too."

Acquisitions and takeovers of textile operations were not uncommon during 1987. New York investor Asher Edelman and Canada's Dominion Textile, Inc., attempted to acquire Burlington Industries. A recent successful acquisition was Walton-Monroe's purchase of Avondale Mills. Other acquisitions included Dan River's Liberty plant and West Point Pepperell's Linndale plant by Greenwood Mills. If ACG were going to put its mill on the market, the time was right.

To try to attach a value to the mill, Boggs called G. Stephen Felker, president of Avondale Mills. Boggs knew Felker fairly well and had known his father for 10 years before he met Stephen. Felker, in his opinion, "seemed like a straight shooter." Boggs was reasonably certain Felker could help him value the plant.

"Only a year or less prior to this he bought Avondale, and Avondale is a denim manufacturer. He had some recent experience determining the value of a denim operation. And he might very well be interested in a high-quality acquisition." It was mid-February when Boggs called him. "Steve," he recalls saying, "this may be the craziest idea I've ever had, but if these farmers who own this denim mill should want to sell it, would you do two things? First of all, would you help me determine what it's worth, and second, would you be interested in buying it?" Felker answered yes to both questions.

Several weeks later, Felker contacted Boggs, telling him he thought the denim mill would be worth between $60 million and $70 million. It was time to take the next step. Felker's estimate was not just a guess. Van May says, "We sent him financial information first off, and I'm sure he looked at it. He knew what the physical assets were; he had just been out here six or seven months before. He could look and see the return on the assets, the kind of profits we were making, and use any number of derivative formulas. For example, if J. P. Stevens was trading on Wall Street at 15 times earnings, and Burlington was trading at 12 times earnings, you could develop a range of what textile companies ought to be worth as a [multiple] of earnings. Then you could look at what the collateral values of the assets really are, what it would cost to replace them. I think that's probably what he did. But he responded very quickly with a price. His estimate of what it might be worth was high enough that we felt like this was an idea worth pursuing. If he had come back with a price that was not that high, we might have just dropped it at that point."

May says the figure Felker proposed "was within the range we expected. It was maybe a little higher than we had hoped for. We were sitting there at the time with ACG on the books for $25 million to $30 million, so I think we felt like if it was $60 million to $75 million, more than double the book value, we had something worth looking at." May ran his own tests, looking at textile companies and using the formulas. He felt the mill ought to "at least be worth $60 million to $75 million."

Boggs approached L. C. Unfred for the first time about the idea. "I remember it very well," Boggs recalls. "I met L. C. at the Lubbock Plaza Hotel. He was en route to the airport to go to a Cotton Incorporated meeting . . . and I met him for lunch. His reaction was about like I thought it would be. He thought I had flipped out. At least he was very cool to the idea." But Boggs, rarely known to improvise, had numbers ready. He showed Unfred, in dollars and cents, what the sale would mean to Unfred and to the other farmers. "At that time," Boggs says, "we estimated that for each dollar of book credits they owned, they would get about $2.35 in cash. They were big numbers. And all the farmers were in bad need of cash at that time. . . . I suggested that L. C. think about it while he was gone, and if he thought it should be pursued to call me and we could set up a board meeting and go from there." Unfred did think about it, and when he returned to Lubbock he told Boggs it should be pursued.

March 26, 1987, was annual meeting day for ACG. It was the most upbeat in the organization's history, surpassing even the celebrations that annual meetings had been in the previous few years. There was nothing to report but problems solved and progress made. No one could have known it would be the last annual meeting in ACG's history.

As was the practice, the annual meeting was held in between meetings of the ACG Board. At the board session that morning, Bob Hale noted ACG had been recognized by LS&CO as its number-one quality supplier for the eleventh consecutive quarter. He also announced his engagement to be married. During the 12 years since the death of his wife and his move to Littlefield, Hale virtually had been married to the ACG mill — 500 employees and 10 acres of machinery. Now, thanks partly to the success of the mill, he was prepared to settle down and share his life with a woman. When he came to ACG, the cooperative surprised him by not yet having the money to build the mill he was supposed to run. Now that he had committed to a lifetime with someone, ACG was about to surprise him again. Later in the day, he would discover that events were unfolding that could see his denim plant sold out from under him, and with it the security he anticipated for his new life. It was not until later that Hale would realize his continued services were considered essential to the consummation of a sale by the plant's prospective buyers. His job never really was threatened, but he had no way of knowing that.

The board recessed, and the thirteenth annual meeting of ACG was convened. L. C. Unfred announced the terms of directors in Districts V and VII were expiring and called for the results of the nominating caucuses in the districts. Not surprisingly, Unfred was nominated from District V and T. W. Stockton from District VII. There were no nominations from the floor, and both men were re-elected to three-year terms by acclamation. The good news poured.

Unfred said the pool versus non-pool controversy had caused problems but appeared to be resolved. Hale reported ACG had opened the most pounds of cotton, packed the most denim, and experienced its best waste savings in its history. Boggs reported new records in the 1985 crop and outturns to 1985-crop members totalling 17.31 cents per pound above the loan. He said another milestone achieved was the retirement of $5.4 million of capital credits to members who had met their advance capital credit requirements. When the annual meeting ended, the ACG Board reconvened. Boggs and May were the only staff members in attendance. Following reelection of officers, Unfred raised the curtain on the surprise of all surprises.

BUT WILL THEY SELL?

The minutes of the meeting show that Unfred broached the subject as though he were stepping through a minefield. He told the board he wanted to discuss a "potential transaction regarding a tentative offer" to purchase the denim plant. He asked Boggs to elaborate. Boggs told the board there was a "good possibility" the mill could be sold for as much as $70 million in cash and he had been asked by the potential buyer not to reveal his name at this time. To say the proposal was unexpected would be an understatement.

Dallas Brewer recalls, "It came all of a sudden. I didn't know anything about it, and I don't think anybody else did, either. Things had been going pretty well for four or five years. We'd been getting a good return from profits from the denim mill." He reflects on his initial reaction. "My first thoughts were, 'Why did C. L. even bother to tell him he would call him back? Why didn't he just tell him no right then?' I didn't want to sell it."

Boggs recalls he had "slides, transparencies, overhead projection, and all this . . . " to explain the idea to the board. The visual aids described the potential transaction and the impact it would have on the members of ACG. He says, "I've never seen such a quiet group during about the 30 or 40 minutes that it took me to go through my presentation. I think at first they were just stunned that I would even come to them with an idea like this. By the time I went through the whole presentation," he says, "I had the numbers and then I had a schedule for each [director] to show them how much cash they would get out of this thing . . . it was pretty impressive. You know," he adds, "you start putting that down to dollars and cents and what it means to an individual's bank account, and then you begin to understand what it's really worth."

Brewer says, "They had my figures there, what I would get out of it personally."

Another director, Van May recalls, stared at the pages in front of him, slowly leafing through them. The man was one whom May thought would have been the most difficult to sell on the idea. He looked at May and asked, "Van, are you saying that if we were able to sell the mill at this price that I would get this much money?" May said, "That's right." The man thought for another few seconds and said, "With this much money I could pay my banker off and tell him to. . . ."

By the end of the meeting, which lasted about two hours, Boggs says "they were pretty warm to the idea of pursuing it." There was a lot of discussion. The board members had just been given a great deal to talk about . . . and to think about.

As the board members were warming to the idea of selling the mill, Boggs jolted them again by suggesting that PCCA might buy it. The directors had begun to see the advantages of selling the mill. But they could not see even a chance that PCCA would want to buy it, or could finance it if it did. Boggs really didn't see it, either. But he felt PCCA should have the opportunity. "It occurred to me," he says, "that if the plant were sold, PCCA and the other regionals which helped ACG get started in the beginning should have an opportunity to buy it. Many of the PCCA members had wanted to get into ACG and were not allowed to do so except through the pool gins. It was important to us," Boggs adds, "that the parties involved in the negotiations all knew exactly where they stood with regard to the sale of the mill. We wanted Avondale to understand that we would like to try to keep the mill in the cooperative family, which seemed a remote possibility at best. If we could not, we stood ready and willing to sell to Avondale if we could agree on a price. That the process be conducted in good faith was our biggest concern." With the board's permission, Boggs notified Felker "that we would continue to negotiate with him in good faith . . . with the full intention of selling the plant to him . . . but that he would have to agree that, once we reached a price, PCCA would have the right of first refusal to buy the plant." Felker was not pleased with the condition. "He agreed to it reluctantly," says Boggs, "but he agreed to it. I really think Stephen felt, as I did, that there was not much chance PCCA could finance the plant even if they wanted to buy it."

A special meeting of the ACG Board was convened late on the afternoon of April 2 at the Lubbock Plaza Hotel. It was an unusual location for an ACG board meeting, but a special guest would join the meeting, and secrecy was a major concern. L. C. Unfred began with a report on a meeting he and Boggs had with Levi Strauss & Co. officials the day before to advise them of the potential sale of the ACG denim plant. To the LS&CO officials, it must have seemed like a bad April Fool's joke. Unfred said, "They were disappointed to learn the plant might be sold, but would support ACG in whatever decision was made." He then revealed the identity of the mystery guest. It was G. Stephen Felker, chairman of Avondale Mills. Before introducing Felker, Boggs said the meeting with him was a "get acquainted" session, not a negotiating session. Avondale had sent a draft contract to ACG and increased its tentative offer to $75 million. Boggs said if the board was willing, it would be possible to complete a letter of intent, but not the final contract, over the coming weekend. Avondale was anxious to get something in writing before word of the possible transaction became public.

Boggs suggested a pool committee meeting for Monday, April 6, 1987, to inform the committeemen before word of a potential sale was received by the membership. Because the members would be the ones who must approve or disapprove a sale, Boggs wanted to ensure they did not hear any wild rumors and that "they were given available and accurate information prior to any vote." He already had informed his senior staff members of a possible sale. Boggs knew that news of a potential sale would spread across the West Texas plains like a prairie fire. And he wanted to avoid any possible overreaction by the local media. Stephen Felker was introduced. He talked about his family's background, their involvement in the textile industry, and the operation of Avondale Mills. Afterward, the group adjourned for dinner. The discussion was neither heavy nor deep. It truly was a "get acquainted" session.

Four days later, on Monday, April 6, Unfred called another special meeting of the board to update it on "current developments concerning the potential sale of the denim plant." Boggs reported that he, Van May, and the association's attorneys had worked through the weekend with representatives of Avondale Mills on an agreement. He said the transaction was very complicated, and many items remained to be negotiated. As an example, he mentioned Avondale's insistence on a cotton purchase agreement as part of the deal. Avondale recognized the importance of

ACG's process of selecting cotton for the mill, so ACG would be required to continue to supply the raw material needs of the mill even after its purchase by Avondale. Boggs said he had not understood this was to be a part of the transaction but Avondale felt very strongly it should be.

Boggs also advised that a letter of intent, non-binding to either ACG or Avondale, had been signed the evening before. It was subject to board approval and was written in order to satisfy Avondale's desire to have "something in writing" before the transaction was discussed with the pool committee. The letter of intent said, "Avondale Mills, Inc. has tentatively offered to purchase the denim plant owned by American Cotton Growers (ACG) located near Littlefield, Texas, including all real estate, equipment, and related fixed assets and intangibles used in connection therewith (the Mill) for a sum of $75,000,000 cash. In addition, Avondale has offered to purchase the inventories, accounts receivable, and other net current assets, less any assumed accounts payable of the Mill for a sum estimated to be $16,500,000 cash.

"The parties," it said, "agree to continue to negotiate in good faith for the purpose of entering into a detailed final binding written agreement . . . [which] . . . must be approved by the respective boards of directors and attorneys of the parties [and], in the case of ACG, the members of said organization. . . ." The only part of the letter of intent binding on the parties was a non-solicitation provision that prohibited ACG from soliciting "any offer from any other person regarding purchase of the Mill." It said unsolicited offers could be considered, as long as the amount of Avondale's offer was not disclosed, with a "prompt and reasonable right of first refusal" in Avondale if better offers were received. It also included a, "non-compete" provision for ACG and an offer of continued employment for Bob Hale. Avondale recognized another key ingredient in the mill's success, Hale's management.

In addition to the requirements stated in the letter, ACG required a fairness opinion of an investment banker as to the worth of the mill, and a 90-day period to reach an agreement of sale, sale and merger, or refinancing, by itself or with affiliated cooperatives, with the right to extend the period for 30 days. Avondale Mills required a due diligence investigation within 45 days regarding: availability of adequate future supply of water; compliance by ACG with various legal requirements, including disposal of hazardous wastes; and absence of any material adverse condition with respect to the mill. ACG would be required to conduct continued operations of the mill in a normal manner until closing and to obtain from its members, as soon as practicable, authority for the board to sell substantially all of the assets of the cooperative.

J. R. Blumrosen, ACG's and PCCA's attorney for many years, noted many instruments would have to be completed with Avondale's lawyers. He emphasized the letter of intent included a "no solicitation" provision and cautioned directors not to divulge any specific contents of the contract with Avondale and not to discuss the transaction outside the membership of ACG. It strains credulity to think that details of the Avondale offer would not go beyond the membership of ACG, particularly when one considers the number of overlapping relationships, not the least of which was the management staff dealing simultaneously with ACG, Avondale, and PCCA. In recognition of this fact, another provision was added to the agreement. It said, "In this connection, ACG shall be responsible only for the use of reasonable means in protecting against a violation hereof; understanding, however, that the amount of AM's [Avondale Mills] offer must be disseminated immediately to numerous key personnel and members of ACG and certain affiliates and that control of the actions of such a large number of people is very difficult if possible at all."

At 12:05 p.m. on April 6, the board adjourned temporarily to attend the textile pool committee meeting. It reconvened at 2:20 p.m. Blumrosen advised the directors that before a special meeting of the members could be called to vote upon any sale of substantially all of the assets of the cooperative, the board itself must adopt a resolution recommending such a sale as required under the Texas Non-profit Corporation Act. The board unanimously approved a motion recommending a sale of substantially all the assets of ACG, including the denim mill at Littlefield. It also voted to submit the motion "to a vote at a meeting of members having voting rights at a special membership meeting on Tuesday, April 21, 1987." The stage was set for one of the most dramatic moments in Texas coop history. The board was asking the members to sell the mill and to permit them to set the terms of the sale. Why? Why would the board even consider selling the "cash cow" that the denim mill had become? To understand, one must hear it from them in their own words.

Clarence Althof had a personal stake in the sale, but adds, "I never was a very large operator, and even with the type of sale we made it didn't amount to a whole lot of money to me. . . . It really wouldn't have made me any difference if we had or hadn't sold.'' He points out that one of his sons, a cotton farmer and ACG member, was opposed to the sale, ''because everything is going fine.'' But he reflects, too, on many other farmers he represented on the board. ''There were two or three years of bad cotton prices, and a lot of the farmers were beginning to get in bad shape.'' He suspects this even included some of his fellow board members. ''With the prospective sale of the mill they had a chance to get back on their feet. . . .''

Dallas Brewer says the proposal kept ''looking more attractive'' in the weeks since it was first aired. ''I was always taught,'' he says, ''if you could double your money on anything it was a good time to sell it. So that's what we finally voted to do.''

Jackie Mull remembers his reaction when C. L. Boggs first told him that ACG was considering selling the mill to Avondale, but that they had arranged for PCCA to have a right of first refusal to purchase it. It was immediately after Boggs and Unfred returned from San Francisco. Mull was incredulous. ''Even if ACG wanted to sell it, why would PCCA want to buy it?'' he asked.

In an April 8 conference call, L. C. Unfred told the board that activities relating to the proposed offer from Avondale were still on track. But, he said, a joint meeting of all the regionals' executive committees could not be held until April 13. Notice of the April 21, special meeting of the ACG membership had been mailed the previous evening, and a news release had been prepared and approved by all parties for release after members had a chance to receive the meeting notice.

Boggs's intuition was working overtime again. He knew that, in a few hours, news of the proposed sale would reach the Lubbock-area media. Some of them might react based on false assumptions, or they might find and interview members hostile to the potential sale. He wanted to prevent that from happening, and the best way was by having the media receive the information immediately after the members. During the conference call, he also suggested it would be advisable to hold five or six area meetings for ACG members prior to the special meeting to discuss the proposed offer to purchase the mill and to answer questions. The board had no objection. In the meantime, Van May was in Dallas interviewing three regional investment banking firms, one of which would be selected to conduct the fairness analysis of Avondale's offer.

Boggs notified the managers of the regional cooperatives that the offer from Avondale included a clause that would permit the regionals to bid on the denim mill. He said they appeared to be receptive to discussing the possibility. It seemed remote. He didn't know it, but a groundswell of sentiment would begin building within ACG's membership to sell the mill to PCCA. Two days later, he told the directors most of the calls the staff had received on the proposal were favorable, although there clearly were some negative feelings toward it. He said many seemed "quite interested" in selling the plant to PCCA, and he added the news release given to the media the previous day was generating a lot of negative conversation. As that was taking place, PCCA's directors and members were themselves looking hard at the proposal and discussing it. It began to develop an air of possibility.

Mull changed his thinking. He says that after "checking it and talking about it . . . it was my idea and my hope that maybe we could buy that mill and put it right into PCCA and operate it so everybody in PCCA, the whole 20,000 members, would share in it."

In the opinion of the ACG staff, the *Lubbock Avalanche-Journal*, with a staff of professional farm editors, reported the issue fairly and accurately. They felt differently about a reporter from one local television station whom they say tried to put a sensational "spin" on the story and ran reports "containing mostly fiction and very little fact" during local news broadcasts. They believe an ACG member opposed to the sale of the mill may have contacted the reporter and monopolized his attention. In spite of the carefully prepared and timed release of information about the proposed sale, the situation was ripe for potentially damaging rumors. The sensationalist television reporter was the worst example. He speculated the mill was going broke and the potential buyer was Japanese — off just slightly.

Fortunately, the area meetings were scheduled to begin that night — Friday, April 10 — to provide information to members and answer questions. The first was held in Brownfield, Texas. The tenor was positive, with active audience participation. The second was held the following afternoon in Littlefield at the Lamb County Agriculture Building, the site of ACG's 1984 annual meeting when LS&CO's Bob Haas had addressed the membership. Again, there was excitement in the air. In fact, there almost was too much excitement when Boggs had a brief confrontation with the hostile TV reporter. Throughout the meeting, members asked direct, significant questions. Some questions, in order to be answered fully, demanded confidential information important to the Avondale negotiations. This placed Boggs in an awkward position. He could not allow his answers to be recorded for all the world on videotape by the television reporter. He turned to the reporter and politely asked him to stop the recorder. At first, the reporter hesitated, as all eyes turned his direction. But the tone of Boggs's voice made it clear the reporter would be asked to leave the meeting if the request were not heeded. He switched off the recorder. He could not afford to be ejected from the meeting and return to the station with a partial report.

On Monday the 13th, the board convened at 10:35 a.m., prior to the joint executive committees' meeting. The directors were told the community meetings in Brownfield and Littlefield, went very well. By this time, many growers were proposing the idea of selling the plant to PCCA. Boggs said an agreement by ACG members to leave part of their proceeds in the company, should they sell it to PCCA, might go a long way in promoting and financing such a plan. He hoped the joint executive committees would agree to reactivate the Special Study Committee and might hire Emmer and Associates to help study the issues, because they already were familiar with both cooperatives' operations.

The board adjourned for the joint executive committees' meeting at 11:45 a.m. The meeting included the PCCA, PCOM, Farmers Cooperative Compress (FCC) and Plainview Cooperative Compress (PCC) executive committees, and the ACG Board. Jackie Mull called the meeting to order and yielded to Unfred, who explained the tentative offer from Avondale Mills. He said the ACG Board had called a membership meeting for April 21, 1987, to consider whether or not to sell the plant, and he noted a letter of intent between ACG and Avondale included a provision for the regional cooperatives to have a 90-day right of first refusal to purchase the mill. Unfred added the joint PCCA-ACG Special Study Committee was still intact and could be called upon to study the possibility of PCCA buying the mill. He suggested PCOM, FCC, and PCC each might want to select three members to serve with this committee, and the group might want to employ an external consultant for assistance. He then called on C. L. Boggs for comments. Boggs said he initially had mixed reactions to selling the plant, but due to the members' need for cash, he had recommended the ACG Board do so. In his opinion, if PCCA could buy the mill, it would be the best situation for all farmers. If this were not possible, ACG would be obligated to sell the mill to Avondale. He then showed slides with detailed information about the proposed sale. Unfred suggested that each regional coop's executive committee caucus after the meeting to select representatives to serve on the new Special Study Committee, and he asked Ed Breihan if he would be willing to chair the group again. Breihan, still recovering from his chairmanship of the last committee, gamely said he would. The joint meeting was adjourned at 2:10 p.m., and the PCCA Executive Committee went into its own session.

The committee members discussed the possibility of PCCA buying the ACG mill, possible financing options open to them, and the likely reaction of the other regional cooperatives. A motion was made by Harold Scaief and seconded by Larry Lockwood (no relation to Lockwood Greene Engineers) to follow Unfred's suggestion and recommend the PCCA Board of Directors reactivate the Special Study Committee, inviting each of the other regionals to appoint three members to serve on the committee. It carried. The executive committee followed Unfred's other suggestion, too, authorizing the Special Study Committee to hire a consultant to assist in the study. Emmer and Associates would again work for the West Texas Cooperatives. In the meantime, the ACG Board itself had gone back into session. C. L. Boggs reported to the ACG Board that the PCCA Executive Committee had adopted the recommendations. Slowly, but surely, the organizations were beginning to work in tandem.

The following day, area meetings were scheduled at St. Lawrence in Glasscock County, and Sweetwater, Texas. Due to the distance from Lubbock, an advance party of staff departed Lubbock early that morning for an afternoon meeting in the St. Lawrence Community Center. Their purpose was to set up equipment for the presentation. To save time travelling between meetings that day, Boggs chartered a plane. The airport at St. Lawrence will never be mistaken for the one at Dallas-Fort Worth. From the air, the narrow blacktop runway was a slit in the prairie. On a spring day in West Texas, the wind and dust make landing at rural airfields an experience that is — "interesting." The flight was rough, but Boggs didn't mind. He was getting a welcome chance to spend some time with his daughter, Rhonda, who had arrived in Lubbock three days earlier from her home in Detroit. She had come to visit her father and had seen little of him because of the mill issue. It was her first private-plane ride, and she gladly accepted his invitation to go along.

The meetings on April 14 were very much like the earlier ones. There was a lot of information given out, and there were a lot of questions asked and answered. The next day, the ACG Board heard a report on the members' response at the area meetings. The directors voted to schedule a meeting at 11:00 a.m. on April 21, just prior to the membership meeting. The staff was asked to draft an appropriate resolution regarding the possibility of extending the sign-in, sign-out period or reopening the pool late should the sale of the denim mill be consummated.

The board met on the morning of April 21, the most significant day in ACG's history. It passed a resolution reopening the pool for sign-ins and sign-outs for a period of at least two weeks if the members approved the sale of the plant. Boggs told the directors he would schedule a news conference to be held after the membership meeting if they did not object. They did not. Meanwhile, the staff was busy setting up computer terminals and audio-visual equipment at the meeting site. At 2:30 p.m., more than 2000 members, staff and associates of American Cotton Growers convened in the Lubbock Municipal Auditorium.

Van May recalls, "We rented the municipal auditorium, and people came over in buses from these communities. They came because they had seen their names on a piece of paper [and they were] saying, 'I'm going to get double my money back in cash on the ACG equity if we sell this . . . I'm going to get $60,000, $150,000. . . .' So they came." May estimates the meeting probably was the largest cooperative meeting ever held in West Texas. Registering the participants was a painstaking process. For legal as well as practical reasons, the procedures had to be perfect.

ACG's lawyers insisted that everyone claiming to be a member had to be verified carefully to ensure he or she was, in fact, a valid member eligible to vote. And with everyone arriving at about the same time, the scene could have been one of chaos. But adequate preparation prevented major problems. More than a dozen staff helped check in the participants. On-line computers were used to check records if there were questions. The registration process showed 1631 eligible members in attendance, and the meeting started only about 30 minutes late. With 4000 eligible members in ACG, 1631 — just over 40 percent — came to the most important meeting the organization had ever called. When asked why so few of the members came, May replies, "You're on the right track, but the question is, 'How did you ever get so many of them to show up?' Our annual meeting at ACG typically would attract less than 100 people." The meeting got underway with L. C. Unfred noting that the cooperative's bylaws and the laws of the State of Texas stipulating the manner in which a membership meeting may be called by the board had been satisfied. A motion was made and seconded to "authorize the sale of all, or substantially all, the property and assets of American Cotton Growers and hereby authorize the board of directors to fix any and all of the terms and conditions thereof and the consideration to be received by American Cotton Growers thereafter."

The vote was the "main event," and the preliminaries didn't take long. Everyone belonging to the cooperative had received material in the mail that not only explained the proposed sale but also detailed their personal stake in it. And as C. L. Boggs noted when he rose to speak, many members had attended one of the six area meetings that had been held in the previous few days. That did not deter him from giving the slide presentation one more time. Van May remembers the presentation took about 45 minutes, and he says it "probably was unnecessary." He heard some comments after the meeting to that effect. Some people had said, "We didn't need to go through all of that," or "We already knew that," or "We came to vote; that's the reason we came." But, May contends, "It was a good way to get the meeting going, and it had to be done." On an issue this important, in an organization based on membership, everybody had to be made to feel that they had been given all the information they needed and every opportunity to express their opinion before taking a vote. There was some discussion, but it was brief. Supporters and opponents of the sale approached microphones placed in the audience. "Several people said 'this is bad,' or 'I don't agree with the numbers you have presented,' or 'you shouldn't sell something that is working for you,'" May recalls. "Others said, 'this is great,' or 'I want this personally,' or 'I recommend that we do it,' and so forth."

Much of the resistance expressed came from young members or those who had signed into the pool for the first time with their 1987 crop. The ACG Board understood these members' concerns; they stood to gain little from the sale. Communicating the benefits of the sale to them would prove difficult. There were, in fact, few benefits for them. Jimmy Nail says the young farmers "came in when [ACG] was on 'a roll.' The mill was running very successfully, making a profit each year. So the young ones didn't want to sell. They were saying that we just told one side of the story, that we didn't tell all the facts. So help me," Nail implores, "we told them everything we knew. I thought we did as fair a job as could be done. But people sometimes just don't want to believe that you're telling them all there is to be known. It's not what they want to hear." The board, as it always had been throughout the history of the cooperative, was frankly more concerned about the well-being of those producers who had been in the pool in the early days, who took the risk, who made the sacrifice, who suffered the loss when things were bad.

Boggs responded methodically to the questions that were raised in the meeting. He had been through this so many times he could have done it in his sleep — when he had time to sleep. The records of the meeting show, during the question-and-answer session, one member thanked the board of directors for its "efforts and many unselfish hours given in behalf of the membership," and the members responded spontaneously with a warm and lengthy round of applause. Then, according to May, as the last ringing of applause echoed faintly through the auditorium, "It seemed to be the general feeling of the group that it was time for a vote." Members called for the question on the motion. Voting was on ballots provided each authorized voting member upon entry into the meeting. The ballots were collected and counted by ACG's auditing firm, Coopers & Lybrand. During the vote counting, members were shown a brief film about the activities of Cotton, Incorporated. Their attention, however, was with the auditors behind closed doors.

After the votes were tabulated, Chairman Unfred called on John Burdette of the auditing firm to report the results of the vote. The auditorium fell hushed. The tension was palpable. To this point no one could possibly know what the outcome would be. The board and staff of ACG were fairly confident the membership would support them. But in the meeting, most of the members had been silent. Those expressing their opinion seemed about equally divided between supporters and opponents. And a two-thirds majority vote was necessary for passage. If only 544 of the 1631 members present voted no or abstained it was dead. Burdette read the tally. Total voting members present: 1631; those not voting: 14; those voting no: 354; those voting yes: 1263. It carried. Burdette announced to the group the two-thirds requirement had been met, and the motion was approved. Van May vividly remembers, "With those words, en masse, the audience stood up and cheered." He adds, "I'm sure the 22 percent who were opposed weren't cheering. But, from the stage where I was, it looked like just a mass rising . . . people going to their feet with their hands in the air, slapping one another on the back, cheering the decision to sell the mill. It rattled the walls.'' He pauses and reflects, ''It was an experience. I'm sure I'll never go through anything like it again.'' The positive, upbeat mood of the meeting was reflected in the news conference that followed it.

Drained of energy and emotion, the members of the board reconvened after the membership meeting. Now they had to consider the inevitable fallout from those who felt the sale of the mill was a mistake. They knew, in particular, there would be dissatisfaction from growers who had gotten out of the program as a result of the chain of problems following the 1976 crop. They also were well aware that producers who had joined the program late — who had only recently begun to reap the benefits of membership and who had not yet built up a sufficient investment in the mill to profit substantially from the sale — would be among the most visible and vocal critics of the move. Dallas Brewer says, ''I believe the people that were more opposed to it than any were the ones who had just gotten into it and had been in it just a year or two . . . maybe younger farmers.'' He supposes, ''. . . they hadn't got their equity built up in it and they wanted to continue to where they could, in 10 or 15 years, have good equity built up. I had one young farmer,'' he recalls, ''tell me that was the only reason he voted against it. He said if he had been in it as long as I had he would have voted for it, too. Another young farmer told him, 'Y'all are the ones who took the chance and stuck your necks out to get it started.''' Across town, at a popular local ''watering hole,'' members of the ACG staff and the auditing firm were unwinding. They were celebrating the end of a grueling few weeks and the successful conclusion of the meeting. Among the patrons was a farmer who had been a member of ACG from the very beginning. He was celebrating, too, because recent investments he had made outside agriculture had gone sour and were threatening his farming operation. The cash he would receive from a sale of the mill would save him and his family. When he noticed the ACG staff across the room, he sent a round of drinks, compliments of a grateful member.

BIDDING WAR

On April 28, C. L. Boggs told the ACG directors a Special Study Committee meeting held the day before had gone well. To those who attended, it was important because formalities were resolved, and deeper concerns were addressed. Compton Cornelius, one of the three ACG representatives on the committee, addressed the group early in the meeting. He explained that the ACG representatives had agreed to step out of any meetings during the discussion of whether PCCA and the affiliated cooperatives should buy the mill. In this way, he explained, any perceived conflict of interest could be avoided. The ACG representatives then excused themselves to allow the committee to discuss their offer. The committee concurred and invited them to rejoin the meeting. The committee felt a need to study the future of denim as part of its mandate and had instructed the consultants to proceed with research. It also decided the that chairmen of the boards and managers of PCCA, PCOM, FCC, and PCC, along with PCCA Vice-Chairman Tex Martin and Jerry Emmer, should meet with Levi Strauss & Co. officials in San Francisco. Jackie Mull asked the committee to invite all members of PCCA's Executive Committee to join the Special Study Committee. He may have suspected that, by the conclusion of the committee's study, there might be little or no time to educate those executive committee members not on the study committee. The committee granted Mull's request.

An old point of contention resurfaced during the April 27 meeting, the issue of merging the regional cooperatives that was discussed a year earlier during the study committee's review of the pool–non-pool controversy. PCC chairman Dolan Fennell stated the committee's study of the mill purchase should not include any attempt to merge the other regionals. Fennell's statement was intended to remind the committee of the compresses' position and the strength of feeling behind it. More important, it would keep the committee's attention focused on the mill purchase. Finally Jerry Emmer outlined an approach for the study and identified critical individual issues that should be investigated. Among these were: distribution of mill margins if PCCA buys the mill, denim market outlook, advantages and disadvantages to PCCA members and sources of financing, condition of the mill's equipment, management and employee status, anticipated capital expenditures, and continuation of a marketing pool. Before the meeting adjourned it was agreed that all PCCA-affiliated gin managers and their directors should be invited to tour the denim mill, and meetings and tours were scheduled for 10:00 a.m. and 3:00 p.m. on April 30, May 1, 5, and 6.

During the ACG Board's conference call on April 28, Van May noted the original letter of intent with Avondale would expire if a definitive agreement were not reached by May 1. He suggested the board extend that deadline to May 15 to allow additional time to get the definitive agreement finalized. A motion to that effect was passed unanimously. On May 14, the board extended the deadline once again, this time to July 20.

The study committee's May 4 meeting at the denim mill was the first attended by representatives of both the Texas Bank for Cooperatives and the Central Bank for Cooperatives. After a slide presentation by Boggs and tour of the facility led by Bob Hale, there were reports on each aspect of the proposed sale. J. R. Blumrosen reviewed the status of the agreement with Avondale; Jerry Emmer reported on the progress of his firm's study; and Van May reported on projected mill operations, offering, "It looks like a good investment for PCCA from a financial standpoint."

By May 22, Emmer was ready to make his final report, most of which was positive for PCCA. The only disturbing news dealt with the financing package offered by the banks for cooperatives. They were requiring a mandatory four-dollar-per-bale retain for every bale of cotton delivered to PCCA. The consensus of the committee members, especially those from PCCA, was that this would cause significant damage to the cooperative. It was another example of the farmers' independence; you do not tell these producers they *must* do something. Emmer also reviewed the need to restructure PCCA's Board of Directors. Mr. Roy's democratic style of management, with 160 members on the cooperative's board, was admirable but totally unsuited to the modern realities of running a professional business organization. The study committee's work was done, and it was time to pass the ball to PCCA's Executive Committee. The group wasted no time, meeting later that same day. But its members had no idea what trouble was brewing.

Earlier in the month, Stephen Felker had started "turning up the burner under the ACG Board," Van May recalls. It began with a disagreement over the date at which the 90-day grace period actually began for PCCA to decide to buy the mill. ACG ultimately lost the argument. Then Felker asked that the grace period be shortened, explaining he needed a decision from PCCA in order to act on another potential mill purchase. The enormous weight of the situation fell squarely on C. L. Boggs's shoulders. Felker was in the driver's seat because the actual contract had not yet been signed by Avondale and ACG. If Boggs and the ACG Board refused to comply, Avondale could withdraw its offer. Boggs, on the other hand, was in a spot because he had told the PCCA Board that they had 90 days to decide on the purchase. Now he had to decide whether to refuse Felker's request and risk losing the Avondale offer or tell PCCA they had less time to make a decision. Neither option was pleasant. He certainly did not want to lose the Avondale offer after the strong mandate from the members to sell the plant, and there was not yet a firm agreement with PCCA. His recommendation to the ACG Board, which it accepted, was that they comply with Avondale's request for the shorter grace period.

Action was needed from PCCA, and it was needed immediately. At the executive committee meeting on the afternoon of May 22, Chairman Jackie Mull stressed the decisions that had to be made regarding purchase of the mill, with financing and board structure the primary issues. It was voted unanimously to recommend to the PCCA Board that:

1. PCCA purchase the textile mill for $75 million plus net current assets of approximately $16.5 million;
2. a one-dollar-per-bale, five-year revolving retain be deducted from each bale delivered to PCCA, as an alternative to the bank's mandatory requirement of four dollars per bale;
3. $12 million subordinated debt be requested from the regional cooperatives, prorated on the approximate financial strength of each regional as follows: Plains Cooperative Oil Mill — six million dollars, Farmers Cooperative Compress — five million dollars, Plainview Cooperative Compress — one million dollars;
4. certain South Texas regionals be invited to participate in the subordinated debt investment in PCCA; and
5. the gins allow PCCA to have 50 cents per bale of their agents' fees.

A discussion about restructuring the PCCA Board boiled over into a lengthy and heated debate. The advocates of reducing the size of the board emphasized that the banks and LS&CO were very concerned about the need for confidentiality, which would be virtually impossible to maintain if the board stayed its present size. Their opponents on the executive committee were few in number, but they were vocal about maintaining a large board, one they felt was closer to and more representative of the membership. A year earlier, this same group had helped lead the opposition to ACG's pool policies. The committee finally agreed to recommend to the PCCA Board that:

1. they recommend to PCCA's membership a change in the bylaws setting the number of PCCA directors at 11 to be elected by districts;

2. the current executive committee serve as this board until the annual membership meeting in August, 1988;
3. prior to that meeting, current districts be realigned to reflect fairly bale volume and members within each;
4. each gin continue to elect a representative to serve on a delegate body to meet regularly and serve as an advisory group to the board.

The motion carried ten to one, an impressive display of unanimity in light of the strong feelings involved.

Six days later, C. L. Boggs had little more than bad news to report to the study committee. The executive committee of Plains Cooperative Oil Mill had voted to reject PCCA's request for subordinated debt, and Farmers Cooperative Compress had yet to act on it. Because each of the regionals' executive committees would be present that afternoon, he hoped that, "some form of compromise might be attained." J. R. Blumrosen then announced the problem was compounded. Avondale had asked the deadline in the letter of intent be moved from July 20 to June 20. It appeared the cards were stacked against PCCA. On the regionals' end, time was being pushed away. On the Avondale end, it was being moved forward.

At the meeting of the regional coops that afternoon, PCOM Chairman Keith Streety explained the oil mill's position on the subordinated debt, FCC Chairman R. D. McCallister confirmed that his cooperative had agreed to delay its decision, and PCC Chairman Dolan Fennell said his committee had not even met on the issue. The meeting recessed to allow the PCCA Executive Committee to caucus and develop an alternative financing proposal. An intense negotiating session ensued between Boggs and TBC's Jack Hughes. Hughes had his limits set by the banks. He said he was authorized to offer a plan including retains of two dollars per bale the first year, three dollars the second, and four dollars the third, plus $10 million in subordinated debt. Eventually, the committee approved the proposal, changing the allocation of subordinated debt to four million dollars for PCOM and five million dollars to FCC, and leaving it at one million dollars for PCC. Mull and Boggs then rejoined the leaders of the other regionals to present the plan. Thirty minutes later, they returned to the PCCA Executive Committee to announce that the plan was unacceptable to the other regionals. PCCA, they had been advised, should, "work out a plan on its own," if possible.

Frustration was mounting. To some members of the PCCA Executive Committee, it must have seemed their fellow farmers, friends, and neighbors were abandoning them. Frustration either leads to resignation or resolve. As was usually the case with the cooperatives, it was the latter. Boggs recommended that: (1) "all PCCA members be given the option each year to participate in the textile mill margins"; (2) "each grower who wishes to exercise such option agree to a five-dollars-per-bale retain on each bale delivered to PCCA for the ensuing three crop years"; and (3) "the textile mill margins be allocated to such members based on that patronage." He knew the damage to be done by any kind of mandatory retain, based on ACG's earlier experiences with the "100 percent concept." He also knew the higher retain would be acceptable, provided farmers had an option of participating or not.

The bank strongly questioned the viability of this "voluntary retain" plan. Hughes explained the ramifications. For the banks to approve it, he explained, PCCA would have to receive commitments from its members for at least 800,000 bales to reach the banks' four million dollars annual requirement. Boggs had enough confidence in the future mill profits that he thought the farmers would participate in order to share in them. And the board had enough confidence that it approved the plan. Hughes said the loan committees of the Texas and Central Banks for Cooperatives would have to approve it, but he said his bank would look on it favorably. The regionals responded favorably, too. And why shouldn't they? They no longer were being asked to accept subordinated debt.

June 2 was to be another milestone. PCCA's Executive Committee met prior to the board meeting to review last-minute developments. The frustrations of May 28 were about to be revisited. C. L. Boggs said the staff had discussed the bank's request that farmers make a three-year commitment to the mill option program, which would allow five dollars per bale to be withheld from their cotton for three years. The staff had concluded that a one-year commitment would be preferable, provided it was acceptable to the bank. He and Mull then went into a meeting with bank representatives. In their absence, the committee approved a plan to establish a marketing pool upon ACG's announcement of intent to liquidate and to accept assignment of ACG's marketing agreements. It also recommended continuation of a premium-for-strength program. As Van May reviewed proposed bylaw amendments to restructure the board, Mull and Boggs returned to the meeting. They brought disappointing news. The Central Bank had rejected the latest financing proposal, and it now was recommending the regional coops guarantee $10 million of the loan instead of investing $10 million in subordinated debt. Alert for opportunities to reduce the cost to the cooperatives, Boggs reacted immediately. Without breaking stride, he pressed for a reduction in the amount of the loan guarantees. The banks said they would consider it, but they never did reduce the amount.

PCOM and FCC soon committed to their shares of the loan guarantee. PCC's Board delayed making a decision on its one million dollar share, and when it finally did meet the sentiment was in opposition. Dan Martin was attending the meeting, and he reminded the PCC Board that PCCA had guaranteed one million dollars of the loan to build the Plainview Compress. This, Martin said, was an ideal opportunity to return the favor. The PCC Board ultimately voted in favor. In the following weeks, Gulf Compress would continue to delay a decision on its share of the loan guarantee. When it voted, it voted "no!" The refusal would not hurt PCCA, but the High Plains cooperative officials were disappointed. Had Gulf's board members already forgotten it was PCCA that kept a cooperative marketing system alive in South Texas when it purchased two small floundering marketing cooperatives there in the late 1960s?

In what seemed an act of futility, the executive committee voted to change the mill option three-year commitment to a one-year commitment. They also recommended the board approve and recommend to the members the proposed bylaw changes on board restructuring. A special membership meeting of PCCA was called for June 23 to vote on the amendments.

Bill Maltby of the Central Bank for Cooperatives attended the June 2, PCCA board meeting. He described negotiations for financing the acquisition and said the bank was very interested in working with PCCA. The acquisition, he told the 60 PCCA directors in attendance, was "financeable," but final details remained to be resolved. Perhaps just as important as these words of reassurance were Maltby's comments about the bank's concern regarding the size and effectiveness of the PCCA Board. After remarks by Boggs regarding the Avondale deadline and the banks' financing requirements, the group was ready to vote. But before they could, a special concern was raised. Director Doug Church of Stanton Coop Gin said he personally favored the acquisition, but he noted members of Rolling Plains Cooperative Compress and Oklahoma Cotton Cooperative Association would be pledging more equity than other members of PCCA. In essence, they were betting their entire warehouse facilities on the future success of the mill. It was a valid concern, but like a voice in a storm, it went nearly unheard. The directors were eager to finish the business at hand.

The motion to purchase the ACG mill carried overwhelmingly. Van May was called upon to review the proposed amendments to the bylaws that would permit restructuring of the board. Quickly now, a motion was made and seconded to recommend PCCA members approve the amendments at a special membership meeting. A roll call vote was requested, and the motion carried, 41 in favor, 15 against, and four abstentions. It is curious that the vote to buy the mill was accepted, but the vote to restructure the board was challenged with a demand for a roll-call vote.

In an anticlimax to what had just taken place, Van May reviewed the executive committee's recommendation to establish a marketing pool and continue the premium-for-strength program, if and when ACG agreed to liquidate. It was approved with little discussion. The PCCA Board then voted to retain 50 cents per bale from local gins' TELCOT agents' fees to bolster PCCA's marketing division. Despite speculation by some cynics, it was not intended to finance purchase of the mill.

Later, the ACG Board would do as PCCA had requested and it had promised, if PCCA were to purchase the mill. It would vote to liquidate the cooperative as soon as all sales proceeds could be distributed and other business concluded. To those present at the day's events, it must have seemed the acquisition was consummated, and the only loose end was the fine print on the loan covenant. Less than two weeks later, that notion would prove a fantasy. The first hint of a snag appeared June 11, when the ACG Board met via conference call. At first, the news seemed good. L. C. Unfred told the directors PCOM had agreed to guarantee five million dollars of PCCA's term loan to purchase the mill. Then he said C. L. Boggs had received a letter from Stephen Felker that asked at what price ACG would sell the mill to Avondale with no right of first refusal by PCCA. What would it take to secure a victory by Avondale? By now, the ACG Board wasn't particularly interested in negotiating further with Felker. It had, from the very start, advised him that no matter what he did, the regional cooperatives of West Texas would have the right of first refusal. The directors had thought a lot about the value of future cash flows from the mill to ACG members and other farmers who chose to continue in PCCA. The consensus was that it would take a large price increase to offset the value of those cash flows. Because a definitive agreement with PCCA was nearing completion, it would have been awkward to back out on the deal with PCCA at that late date. But the ACG Board had no idea how much Felker wanted their denim mill.

Four days later, on June 15, the PCCA Executive Committee met via telephone at seven o'clock in the morning. Boggs reported to the committee members on the status of the tentative contract with ACG; then he read the letter Felker had written to the ACG Board with his offer to raise the stakes. The ACG Board, he told them, had reconfirmed its desire to sell the mill to PCCA if it substantially matched Felker's offer. The ACG Board met that same afternoon. Boggs said the PCCA Executive Committee was scheduled to meet the next day to sign the contract with ACG. The news was bolstered by Van May's report that PCCA had received a commitment for financing from the banks for cooperatives. All the pieces were nicely in place. On the next day, it would be over; the long march would come to an end. And the ACG Board recessed for the night.

Within hours, all hell broke loose. Boggs placed a telephone call to Stephen Felker to tell him that ACG felt obligated to pursue the PCCA offer, because negotiations had reached such a late stage. Felker's vague offer, his "what price?" approach, was not of interest to ACG. Twenty minutes after the conversation, Boggs's telephone rang. It was Stephen Felker. He was raising his bid for the mill to $85 million, an increase of $10 million. Boggs was stunned. He could not believe what he was hearing. Everything had been painstakingly put in place — the support of the cooperatives and their members, the plans, the financing — everything. This could blow it apart. Boggs told Felker he would talk with the ACG directors as soon as possible and let him know their response. Late at night, Boggs placed a call of his own. He called Van May at home to tell him what had happened. May wasn't surprised to get the call; Boggs was known for telephone calls to his staff at odd hours. But May was surprised by what Boggs told him about Felker's call. Then he remembers Boggs saying, "I know you can't do anything about this, but I wanted someone besides myself to be worrying about it." Boggs then called L. C. Unfred, told him about the offer, and got his approval to schedule a conference call with the ACG Board for the following morning. That day, June 16, 1987, would be one that would tax the patience, determination, and endurance of everyone involved in the proposed sale of the mill.

Climax

It began at 6:55 a.m., when Boggs notified the ACG directors in a conference call. The board talked about the value of the additional offer versus the value of future cash flows to members if the mill sold to PCCA. There was concern among at least some of the directors that Avondale's offer might not be binding. ACG's attorney, Joe Boerner, of Crenshaw, Dupree and Milam, hired to assist in the sale of the mill, recommended the directors require written notice of Felker's latest offer before continuing negotiations with PCCA. He also advised the board to notify PCCA that ACG, in all likelihood, would not be in a position to make a final agreement that day. The conference call meeting adjourned at 7:35 a.m.

At 8:07 a.m., Boggs called Felker and asked him to make his "final and best offer" in writing and to telecopy it to him no later than 11:00 a.m., because ACG and PCCA were scheduled to meet and finalize the mill sale. Though he tried not to let it show in his voice, Boggs was getting tired of the roller coaster his emotions had been riding.

At 9:30 a.m., the PCCA Executive Committee convened in its conference room. Boggs reported Avondale's offer and said the ACG Board had asked Felker to put his latest offer in writing by 11:00 a.m. His voice showing fatigue and concern, he said he hoped they could, "keep the mill in the coop family." The executive committee voted to accept the proposed contract with ACG and offer a firm price of $75 million for the mill. After that, there was nothing to do but watch the clock and wait for Felker's next move.

At 10:05 a.m., Felker called Boggs and said he could offer more than $85 million, but he was not sure how much more. He was in a hotel in Oshkosh, Wisconsin, that morning and wanted to return to Avondale's headquarters in Alabama to work on the proposal. Boggs insisted on receiving the written offer by 11:00 a.m.

At 11:15, Felker called again and said he had found a member of the hotel staff to type the document and telecopy it to Lubbock. When the offer arrived, it stunned the cotton growers. This time he was offering $90 million, a $15 million dollar increase from his original offer. He called again at 11:40 from a pay telephone to make certain they had received his offer. Since Boggs had told them of the $85 million offer at 9:30 that morning, PCCA Vice-Chairman Tex Martin and his colleagues were deeply concerned and doubtful of the outcome. They had recessed for lunch when the latest offer from Felker came in.

At 12:40 p.m., the PCCA Executive Committee reconvened. Boggs showed them the $90 million offer from Avondale. As soon as it was announced, Martin was heard to say, "That's it, then." He had lost all hope of PCCA acquiring the mill. Although the reactions of the committeemen were not as visible, it is certain they shared his feelings of resignation. All that time. All that work. All that emotion. For nothing. The PCCA Executive Committee members discussed the matter briefly. Then, at 12:50 p.m., they went to their board room to resume meeting with the ACG Board.

It was a brief meeting. There wasn't much to say. When it was over, the executive committee members rose to leave the room. As they were on their way out the door, they were stopped by the words of L. C. Unfred. The decision over Avondale's move, he said, would be pure agony for the ACG Board. More than anything, Unfred and the rest of the board wanted PCCA to be able to buy the mill. The committee gathered in the hallway outside the board room, "ready to throw in the towel," Van May recalls. "Their comments were, 'We can't do that,' and 'We can't do $90 million.'" Some of them began walking toward the door. Better to go home and tend to farming than stand in a building and mourn the death of something that wasn't meant to be.

C. L. Boggs remembers that about that time May spoke up, and everyone stopped to listen. May recalls, "I asked the committee to consider making some kind of counter-offer, anything they could. I suggested this would leave a crack in the door and at least give ACG something to talk about. It would keep the dialogue going." The executive committee huddled again and began to consider again. They were farmers, damn it! And cooperative farmers, too. Every day of their lives presented challenges. Life itself was a struggle. But they didn't give up. They worked together. That's what had saved them in the past, and that's what would save them now. Those drifting toward the exits stopped and turned and listened. And slowly, they came back together. Only minutes after conceding defeat to Avondale, the committee was together again, talking again, focusing its collective energy on finding a way to win.

FAMILY VICTORY

It was not easy. Several members of the PCCA Executive Committee felt the transaction was, in the words of Van May, "too big a piece to swallow." PCCA would be using its $28 million balance sheet to leverage a deal nearing $100 million. They were well aware, says May, "that if something happened along the road, they could end up losing PCCA.'' They couldn't match Felker dollar for dollar, but in the end they decided to come as close as they could.

Minutes after their meeting, PCCA Chairman Jackie Mull went before the ACG directors once again. He told them PCCA would offer ACG a price of $82.5 million for the mill, plus an estimated $16.5 million for net current assets. Furthermore, he continued, PCCA would buy ACG's stock in the oil mill and the bank for cooperatives for an estimated net present value of just over two million dollars. He said the offer was unanimously supported by his committee members. And he cautioned, as everyone was painfully aware, the offer would require formal board and bank approval. Now it was ACG's turn to caucus in private. The directors, knowing fully the importance of the bank's role in the transaction, directed Boggs and May to obtain permission from the PCCA Executive Committee to contact the bank immediately regarding the necessary additional financing. The joint meeting recessed until word was received from the bank at 6:45 p.m. Then the ACG Board reconvened by itself.

Boggs said the bank had just informed him it would be unable to give PCCA an answer until 11:00 a.m. the next day. By now, exhaustion was beginning to take its toll and confusion beginning to show. Attorney Boerner reminded the directors their duty was to weigh all the factors and consider the best interests of the ACG membership as a whole — to use their business judgment and good faith to make any decisions. Discussion centered on why ACG should accept a lower price from PCCA. The answer was that ACG members would then be able to continue to participate in the mill ownership and earnings through PCCA. Van May recalls a brief but eloquent speech by Jim Bob Curry. It was perhaps less than a minute in length, but May says it was a perfect summation of why the mill should be kept in the coop family. After a long discussion, Harold Barrett moved, and Doyce Middlebrook seconded, to accept PCCA's offer pending approval by the bank and PCCA's Board of Directors, and to execute the final asset purchase agreement. The motion carried five to one.

It was T. W. Stockton who cast the lone dissenting vote. He makes it very clear that he was not opposed to selling the mill. ''I was against selling it for less money than Avondale offered. At that time,'' he says, ''many of the young farmers in my community were against selling the plant to *anyone*. I just didn't think we should sell it for less than the Avondale offer.'' Like Stockton, all the directors wanted the best price they could get, but most did not think that PCCA could afford to pay that much for the mill, even if it wanted to. They were willing to take less in order to keep the mill a cooperative enterprise. Stockton, asked today how he views the issue in retrospect, has no doubts. "We did the right thing."

Did ACG get enough money for the mill? Did PCCA pay too much? The answers to those questions depend squarely upon whom you ask. Kervin Frysak, a young ACG producer from Midkiff Coop, has no doubt that the mill was sold for far less than its value. "Even during the Depression," he says, "any type of business that was selling was selling for 15 to 20 times earnings. Well, the denim mill was making $17.5 million. Twenty times that . . . ," he pauses to let the size of the numbers sink in.

To this day, Jackie Mull feels the price was too high. "I think the first price of $75 million would probably have been a fair price," he says. "I understand the ACG Board's position. They wanted to sell the mill to PCCA, but they had a $90 million offer from Avondale they couldn't afford to ignore." Mull continues, "There were some who questioned whether or not the $90 million offer from Avondale was legitimate." There certainly were, and they question it to this day. Some in Midkiff are cynical enough to believe that Avondale was only a "straw man" to goad PCCA into buying the mill. Rocky King, the gin manager there, says Avondale was nothing more than a "vehicle to get [the mill] moved from ACG to PCCA."

Billy Eggemeyer, a producer with a reputation for saying what is on his mind, feels certain that ". . . management went and contacted Avondale Mills. [Management said], 'We need an offer so PCCA can buy this mill.'" Eggemeyer considers the increased offer from Avondale a ploy to push the sale. He calls it ". . . B.S., because as soon as we got in there, at this meeting, [the staff said] 'Hey, man, they called and offered us $15 million more for this mill! What are we going to do now?'"

To the doubters, Mull replies, "I don't have any doubt because I saw the letter of intent, I saw their attorneys, ACG attorneys, and PCCA attorneys working on the project." He says he has no doubt that ACG would have sold the mill to Avondale for $90 million if PCCA had not raised its offer to $82.5 million.

After the ACG Board rejoined the PCCA Executive Committee, a question was raised as to whether the two cooperatives should sign a letter of intent that evening. Again, the ACG Board excused itself to discuss the question. Boerner said the attorneys could draft some type of commitment letter if the board desired, and the board did. The letter would be written and signed that night, though it all was contingent on approval by the bank and the PCCA Board. With that decision, the meeting was adjourned at 8:40 p.m. It was time to drive home (for some, an hour or two over country roads), get a few hours sleep, and prepare mentally and physically for whatever the next day would bring. At this point, no one could even guess at what the next day would bring. For Unfred, Mull, the staff, and attorneys, the day would go on another two hours. That was how long it took to draft the letter for the chairmen's signatures. It had been an incredible day. At the end of it all, two men, each dedicated to his cooperative and determined to do right for it, sat down and put their names to paper.

The next morning, according to Van May, "the bank called back and said, 'Well, there are two or three things that have to be guaranteed before we can do this.' C. L. said to the bankers, 'Look! You guys are not getting the picture. We can't guarantee *anything* between now and noon today!' What the bank had in mind was for the other cooperatives to guarantee an additional one million dollars of the debt. We couldn't guarantee that. C. L. told them, 'If you make that a condition of the financing right now, and we can't meet that by noon today — and we can't because these other cooperatives' boards would have to get together, and they can't — then you're saying that at noon today this mill goes to Avondale. That's what you're saying if you are going to make us do that today. . . . If you want to see this business leave control of the cooperative and the cotton farmers of West Texas, then all you have to do is make that a requirement of the loan.'" Ultimately, the bank relented.

At 12:35 that afternoon, the executive committee met again, this time by conference call. Boggs reported that the banks for cooperatives had agreed to finance PCCA's request for an additional $7.5 million in term debt, with conditions that:

1. the additional $7.5 million be priced at a premium of approximately 150 points (1.5%)over the negotiated interest rate on the $75 million loan;
2. the additional $7.5 million not be amortized until the latter of June 30, 1990, or the attainment of a one-to-one long-term debt-to-equity ratio, and the additional amount be amortized over the remaining life of the larger loan unless staff negotiated a shorter term;
3. the guarantee from the other cooperatives be increased by $1 million, if possible;
4. the per-unit retains be increased to total $22 million over the next five years instead of $20 million originally proposed; and

5. an origination fee of 1/4 of one percent, approximating $206,000, be paid to the Bank for Cooperatives.

Boggs said he and the staff had agreed to the conditions, although they did not like them.

June 19 was filled with anticipation for the ACG Board, the PCCA Executive Committee and the staff. Jackie Mull called the PCCA Board meeting to order at 11:00 a.m. This meeting had been scheduled several days earlier to vote on the $75 million purchase price. The board registered surprise when told of the increase to $82.5 million, but the news didn't sour them. Van May read the resolution unanimously approved by the executive committee and, after much discussion, the question was called and a roll-call vote requested. The resolution passed, 38 to 9. Weeks of sweat, anguish, and sleepless nights for directors and staff were ended. The family was whole again.

Of the PCCA board members who opposed the purchase, Mull says they were "just kind of like I was at first. [They were asking] 'Why would we want to buy it, even if they want to sell it?' Eighty-two-and-a-half million dollars is a lot of money, and they didn't understand how we could pay it back." He says a lot of PCCA members who were not on the board probably were against buying the mill, "because they hadn't been through the [Emmer] study, hadn't seen it, and didn't know what was going on." Typical of their mutual respect, Mull and Unfred read statements commending each other's cooperative. Unfred thanked Mull and PCCA for making the decision that would keep the mill "in the cooperative family."

Later that afternoon, Unfred and Mull signed the contract effectuating the sale of the denim mill and related assets to PCCA, and Mull presented Unfred with a check for more than $4 million in earnest money. The pressure lifted, a few ACG directors lounged on a sofa outside C. L. Boggs's office. They alternately joked a bit with the senior managers' assistants and just sat silent, calm, content.

The PCCA membership meeting on June 23 was almost anticlimactic. One issue — the restructuring of the PCCA Board — remained to be considered. Members squeezed into the cooperative's board room to hear their new "partner," Levi Strauss & Co.'s Chief Executive Officer, Bob Haas. Haas, attired in a suit made of Littlefield denim, offered remarks that were brief but stirring. He talked about the history of his company — his family's company — and its relationship with ACG. Most of all, he expressed his and LS&CO's happiness with the prospect of a continued relationship with the farmers of Texas and Oklahoma. Haas subtly reminded the members of the need for confidentiality between their cooperative and his company, explaining the intensely competitive nature of the apparel industry. The members heard the message directly from the man outside the cooperative who was most important to its continued success. If any thoughts that the need to restructure PCCA's Board was a scheme or a figment of someone's imagination, Haas's talk should have been erased right there. Soon after Haas's remarks, the members voted overwhelmingly to approve the amendments to their cooperative's bylaws and reduce the size of the board to 11 directors.

Later that summer, PCCA opened its first marketing pool to West Texas and Oklahoma members. ACG members automatically became members of the PCCA pool, their marketing agreements having been transferred. If they wished to sign out they could do so during the month of August as interested PCCA members were signing in. The overall response to PCCA's initial pool exceeded the most optimistic expectations. South Texas members of PCCA would have to wait until the following year to participate in pool marketing. Late June was only a few weeks from the South Texas harvest and much too late to organize a pool. In February, 1989, a separate South Texas pool was opened, and again response exceeded projections. The reasons for a separate pool in South Texas were clear. The harvest period occurs earlier and a higher quality of cotton is often produced in that region.

In 1987, for the first time, all PCCA members were given the opportunity to invest in textile manufacturing through their cooperative, and their response was indicative of the directors' sentiments during the mill acquisition period. Members committed more than 850,000 bales ($4.2 million retained) that year, greatly surpassing the banks for cooperatives' 600,000-bale, three million dollar requirement. The 1988 crop mill option participation increased to over one million bales in the mill option, totaling over five million dollars retained, again exceeding the banks' requirements. Nature contributed significantly during the 1987 and 1988 crops, allowing Texas and Oklahoma growers to harvest consecutive record yields. Maybe it was good luck for PCCA. The result was that PCCA's venture into textile manufacturing was off to a better start than most would have dared dream.

By August 4, 1987, auditors had completed their review of the mill's inventories and assets. The paperwork was ready as L. C. Unfred and Jackie Mull met again to finalize the change in ownership.

That day, ACG paid its loan in full to the Texas Bank for Cooperatives, and PCCA assumed a greater debt than it had ever done in its 34-year history. PCCA accepted the challenge, knowing it had the support of its membership and a proven winner in the denim mill. Earlier, on July 30, PCCA made its final note payment to the TBC. For four short days, PCCA was term-debt free; for the first time in over 30 years, it carried no long-term debt. Then it borrowed $82.5 million to purchase the mill.

On October 21, ACG held a membership meeting. Where 1631 members had attended the meeting at the auditorium, 56 voting members were on the registration list of those attending this meeting. Fifty-six members were on hand to ring down the curtain. L. C. Unfred noted that the bylaw requirements and the laws of the state of Texas stipulating the manner in which a membership meeting may be called by the board had been satisfied. Minutes of the annual membership meeting held March 26, 1987, and the special membership meeting held April 21, 1987, were approved as read. Unfred asked all ACG members to stand and be counted in the event some members failed to register properly prior to the meeting. He said the purpose of the meeting was to vote on the dissolution of ACG. He then called on Van May to read a resolution, copies of which had been distributed to those members present, calling for the dissolution of American Cotton Growers. He noted that the board of directors recommended this action to the members. Following this reading, a motion was duly made and seconded to adopt the resolution. Members voting for the resolution were asked to stand, and the resolution was approved unanimously. C. L. Boggs reviewed the PCCA Board's unanimous action to operate the denim mill as "American Cotton Growers, A Division of Plains Cotton Cooperative Association." L. C. Unfred said the dissolution of ACG caused mixed emotions and called this the beginning of a new era. He introduced each member of the ACG Board of Directors. Each was warmly thanked for his efforts with a round of applause. There being no further business, the meeting was adjourned at 2:40 p.m.

Looking Back

How successful was ACG? In 1975, ACG was the Crosbyton Gin Division with a net worth of approximately $2.6 million. Twenty-seven cooperative gins and the Lubbock regional cooperatives invested $5.9 million in ACG, then borrowed enough money from the Bank for Cooperatives to build the denim mill. Since that time, the Crosbyton Gin Division has been spun off and all its original investment taken out of ACG. All the cash invested by the gins and regional cooperatives has been repaid with interest. In all except one year (the 1976 crop), ACG members received substantially more for their cotton than did other cotton growers in West Texas. When the plant was sold, over $115 million was distributed to these ACG members. This represents a huge return on investment, one of which few businesses in America could boast.

The money each ACG member received from the sale of the mill was directly related to patronage, the number of bales of cotton that he or she delivered to the program throughout its life. Being a director of ACG did not pose an advantage. Although most of the ACG directors are substantial farmers, there were members of the association who are much larger farmers than the directors, and who fared considerably better. C. L. Boggs recalls the largest single amount received by any one individual "was over $900,000. There probably were families that had several ACG members who together received more than that. But it was not unusual for individuals to get $50,000, $100,000 or $150,000. And that," he adds, "is why the sale took place. I think every single member of ACG," Boggs says, "thought selling the plant was a crazy idea when they first heard it . . . wild to sell something that had been this successful. Basically, farmers and coops don't sell things that are successful. But when you put it in terms of dollars and cents . . . and how much it meant in relationship to how badly they needed cash in their farming operations, they decided that they needed the cash more than they needed the denim plant.''

Boggs reflects on the minority opinion. ''We had a lot of members of ACG who thought we made a horrible mistake selling the plant. Some feel that way to this day.'' He contends that most came out of the transaction as winners. ''The ACG farmers,'' he says, ''who built this plant and were in the program in the early years, made a heavy commitment. I don't mean in cash, because the only cash commitments the farmers made were through their coop gins who invested the $10 per bale, and the $3.5 million invested by the regional coops. The regionals' investment was a risk taken by all the farmers who owned them. But the big risk was taken by the farmers who pledged their cotton to the pool as part of the collateral used to borrow the money to build the plant.''

He uses an example of a farmer who was an original member of ACG and had 400 bales of cotton each year. ''He had to commit in April of 1975, and every April thereafter if he stayed in the program, that he would deliver all his cotton to the coop. He did not know how much money he would be paid, only that it would not be less than the government loan rate. In many cases,'' Boggs says, ''back in 1975, 1976, and 1977, cotton might have been worth $50 to $100 per bale more than the CCC [Commodity Credit Corporation] loan rate, and he was being advanced the loan rate. If the denim mill had failed, and if ACG had gone bankrupt, this grower would have lost all of that excess value for his cotton. If the value of the cotton was $100 per bale more than the loan, he would have lost $40,000 on his 400 bales. Now that,'' he says, ''was a big risk.'' In fact, it actually happened on the 1976 crop, when ACG had to keep $30 per bale value of the lint to make its note payment. ''When the plant was sold,'' Boggs continues, ''and these members got over $2.50 for each $1.00 in book credits they owned, they were getting their due. ''They took a big risk, they weathered the storm during the bad years, and now they were finally getting a return on their investment. They clearly were winners.

''Some of my most rewarding moments in working for the coops,'' he says, ''have been when people tell me how much the money they received from the plant sale meant to them. One large grower whom I have known for years told me he was planning to file bankruptcy during the summer of 1987. The money from the sale of the mill saved him. Another member told me he was going to pay off his banker for the first time in years. He said, 'When I pay him off, I'm going to tell him where to put the money,' though I don't think he ever really did. One member said he bought a new Buick and it was colored 'denim blue.' Another said he would 'build Mama that new house she has been wanting for years.' I had seen many farmers whom I considered to be good operators go from being in good financial shape to being in serious trouble," Boggs adds. "The proceeds from this sale got many of them back on their feet again. It is a good feeling to have had a hand in helping people like that. So, when I'm asked if I ever had second thoughts about selling the denim mill, my answer is an emphatic 'No!'"

But the ACG members are not the only winners Boggs sees in the transaction. "It was a win situation for PCCA and the farmers who are members of it. They weren't taking the big risk that was taken in the early years. But they were able to get into the textile business by paying a little higher price for the mill. They are getting to buy an ongoing business, with an established track record, with a customer [Levi Strauss & Co.] that is the strongest in the world."

"The customer" looks back on its experience with ACG. Tom Kasten, now president of Levi Strauss & Co.'s Women's Wear Division, provides an observation that helps explain the frequent use of the word "partners" to describe the relationship. "One of the more difficult tasks between a buyer and a seller is negotiating," Kasten says. "[Negotiations] often are tests of strength, stamina, wills, and egos. But never did I have an acrimonious or unpleasant negotiation with ACG. Part of this," he adds, "may be due to the formula nature of price determination [in textiles], but still there were times when denim prices were either sky-high or severely depressed that could easily lead to one side to want to take advantage of the situation." That never happened between LS&CO and ACG. Of ACG, Kasten says, "They always conducted themselves as gentlemen."

Bob Haas still is amazed by the phenomenon of American Cotton Growers. He likens ACG's success to that of an "NFL expansion franchise that was in the Super Bowl in its fourth year of existence. These people came from nowhere," he says, "and topped people who were consistent Super Bowl performers." Haas calls the denim mill, "world class. When I say world class, I mean anyone who can be consistently number one in the Levi Strauss quality rankings automatically is at the very upper tier of fabric producers in the world. To have accomplished that much in such a short period of time in competition with mills that have been weaving denim for decades is an astonishing tribute to Bob's [Hale] leadership and technical skills. [It also is] a tribute to the courage of the ACG membership and board of directors in making the investments in modern equipment and just their grit and fortitude to be the best. It is,'' he says, ''remarkable.'' Haas is well aware there was resentment on the part of a few other suppliers when ACG was buffered to some extent from market downturns as a result of its contract with LS&CO. ''That is understandable,'' he says, ''[and] would have been even more intense on the part of those mills who no longer do business with Levi's, since we have narrowed the number of denim suppliers. Their presumption is that because we had these secure relationships they were being cut back. In point of fact,'' he adds, ''the reason that we discriminated was that our remaining suppliers simply did the best job in terms of quality, service, and price. So it was based on performance. I think, at the same time, there is a certain degree of envy and respect that has to be accorded to a mill that came from nowhere and consistently topped the quality rankings.'' He describes the relationship between his firm and ACG as, ''either very old-fashioned or very modern. It is old-fashioned in . . . that it was based on a handshake and an understanding between people of integrity on both sides. [It is] highly personal rather than institutional, because it is based on integrity and trust and an understanding of each other's needs. I would say it also is a highly modern relationship,'' he adds, ''in that . . . more and more . . . enterprises are entering into what I would call strategic alliances . . . and clearly ACG's being willing to devote all their output to Levi's and Levi's being willing to commit to all the production of ACG . . . is an example of a strategic alliance.'' The relationship, he adds, ''exemplifies the best in what we hope for in partners.''

Chairman Peter Haas, Sr., calls the relationship with ACG, ''one of the high spots of being in our business, to have a relationship as close and firm and trusting as this is. I think,'' he adds, ''this is the way all business relationships should be.''

C. L. Boggs talks about the synergy created by the merger of the two cooperatives. ''With the members of ACG joining the PCCA group, we now have over 20,000 members in PCCA. We [PCCA] market more than 95 percent of the cotton grown by these farmers, over 1.5 million bales in a normal year, with a gross dollar value of $250 to $400 million. I can't say enough either,'' adds Boggs, ''about the financial stability the denim plant has brought to PCCA. We have been fortunate to have two good years since the plant purchase. The denim mill contributed $10 million to PCCA's earnings the first year and expected profits are over $4.5 million this year, which happens to be a year of very poor denim prices. PCCA's net worth now is over $50 million and growing.'' The $50 million net worth is almost 4200 times the $12,000 initial investment in 1953, when Roy Davis used that much in oil mill checks to establish PCCA. ''With annual sales of $300 to $400 million and diversified profit centers,'' Boggs says, ''including the denim mill, PCCA is now in its best position ever to serve its cotton producer members of Texas and Oklahoma.'' But to Boggs, the most important result of the mill sale is that ''we were able to eliminate the greatest controversy that's been in the coop family in Texas and Oklahoma, probably ever.

''We were about at the end of our road. I have enjoyed my 23 years of working for the coops, but I will have to say there was not much to enjoy during the two or three years leading up to the sale. The pool–non-pool controversy was out of control. I was caught in the middle as the manager of both coops, as was the rest of the staff, and there seemed to be no solution. Although we were doing everything we could to be fair and impartial, we were the only common denominator. I was determined not to let the two coops split if it could be avoided. In my opinion, it would have left scars that might never have healed. I have seen communities split over an issue, and one group would go a short distance, perhaps across the street, and build another coop gin. Many years after most people had forgotten what the original split was all about, assuming the two gins could survive economically in competition with one another, there still are hard feelings. This would have been magnified many times if PCCA and ACG had split. The repercussions would have carried over into the other coops, and the effects would have been devastating.'' Boggs continues, ''Although I was determined not to let it happen if it possibly could be avoided, I was about to give up.'' He describes the feeling as being like ''on a merry-go-round as a kid. It was going much too fast, and I wanted to get off but I was afraid to turn loose.

''As I got over the shock of Van's suggestion that we sell the plant and thought through the ramifications, I became more excited. When you combine the benefits of restoring harmony with the amount of sorely needed money the ACG members would get out of the sale, it seemed like an excellent solution to multiple problems. With PCCA buying it, we have the best of all worlds. The harmony that returned to the coop family has exceeded our expectations. I'm certainly not suggesting that all the past has been forgotten and everyone is 'living happily ever after.' I know there still are some hard feelings, but they will pass. If a movie script had been written, I don't know how it could have ended much better than it did.''

The new environment has made life easier for the PCCA staff. Boggs says, ''Someone once said, 'I didn't know how bad I hurt until I felt good again.' We can relate to that statement. Since PCCA bought the denim mill, we feel like we have launched new careers. It is fun to come to work again.'' In a personal postscript, he adds, ''I haven't had a bad 'chewing out' in almost two years.'' The outcome, he adds, is ''a stronger coop than we ever had before.''

Dallas Brewer also feels the outcome was worth any negative reaction. ''The last four or five years, we'd had it pretty rough in farming, prices down and costing more to farm, and we had some heavy debts. A lot of people were having to file bankruptcy because they just couldn't get financing. They had gotten to the point that they owed more than they could ever pay off. And this sale of the denim mill,'' Brewer asserts, ''saved a bunch of them from having to declare bankruptcy. And, of course, for the older people who were getting close to retirement, it was just a dream come true.''

Clyde Crausbay, one of the original Crosbyton board members and now vice-president of the Associated Cotton Growers Board, voted for the sale, ''reluctantly.'' He had left the pool in 1979, several years after the plant's cost overruns and other problems, and then came back in 1984. Why would anyone leave the pool at the onset of good times and so long after the bad times? His response is typical of the long, slow, and deep reflection of a man who builds his life around the growth of crops. ''The cost overrun . . . was still bothering me,'' he says. The expression on his face says it does to this day. But his decision bothers him, too. Because he left the pool for five years, Clyde Crausbay did not stand to gain as much from the sale of the mill as he would have had he stayed all the way through. Now he says, ''I guess I was taking the easy way out. I did all right [on the sale] but not near what I would have.'' His remarks lend added insight to the expectations of the participants at the beginning of the project. ''At the time,'' he says, ''the stock didn't mean anything to me because we were talking about retiring it in a number of years and I wasn't figuring on getting any. "Nobody ever heard of a place . . . a coop . . . selling out for twice their value," he adds. "I've heard of them going broke. I've heard of them merging. And I've heard of everything but selling out and paying two-and-a-half times." Crausbay's original response to the proposed sale of the mill had been, "No, I'm not in favor of it; we're just now getting to make money." He later changed his mind. "The rest of them were [in favor] and there's a lot of people that maybe saved their farming operation [through the sale]. So I decided to vote for it." He adds several other reasons for his decision. One is the settlement he received. And the other, possibly even more compelling, is "My wife said vote for it."

When asked if he voted for the sale, Wayne Huffaker, candid as ever, replies completely in character, "Oh, yeah!" Why? "Because I needed the money desperately. I'm busted!" On a less personal level, he adds that the sale was "the only way the controversy between the two cooperatives was ever going to be put to rest. It was, in my opinion, about to break the cooperative system completely apart. It would have been a disaster."

Mondell Mills, manager of the Crosbyton Gin, had no stake at all in the sale of the mill. "From one point of view," he says, "I hated to see it happen. . . . We had this thing going that our competition [the non-pool and independent gins] didn't have. And everybody wants something their competition doesn't have. But when I sat down [in the municipal auditorium] and I looked at the little, gray-haired ladies — and I saw several of them at the stockholders' meeting to vote 'yes,' they wanted to sell it — they were going to get $25,000 or $30,000 and that was more money than they had ever had in their hands at one time." He looks incredulous. "How in the world could a man be against a deal like that? Just not any way." As to the effect of the sale on the gin, Mills harbors no misgivings. "I have more confidence in our membership than to feel our gin would be hurt. I felt they would continue to be loyal and it would work for us."

When it was announced the denim mill would be purchased by PCCA, the reaction at the mill could best be described as a collective sigh of relief. Uncertainty breeds concern, and Danny Davis says the plant staff was deeply concerned about the mill being sold to a textile company. Autonomy, or the loss of it, was the major factor on their minds. Not that the mill ever was completely autonomous. "We report to the farmers," Davis says. "But we have the ability here at this plant to make adjustments in our operations to fit what best suits our customers without having to go through a corporate headquarters. We would have lost our identity.'' Asked if he could think of any advantages that might have accrued to him and the rest of the mill staff by becoming part of a textile concern, he says ''It might have opened up some promotion opportunities for some of us. If you did a good job here they might offer to move you somewhere else in the company with a promotion and a pay raise. Promotions don't usually come as fast when you have a one-plant operation as when you have a multi-plant operation.'' And, he admits, ''We'd like to think we know everything about making denim, but there are probably people who know things we don't know. We might have learned some things from them. But eventually,'' he adds, ''you learn those things anyway.'' As for the ''new'' owners, Davis says ''There haven't been many changes made at all. The good folks over at Lubbock [the PCCA Board, owners, and staff] have been easy to work with and, in my experience with them, they don't try to interfere in areas where we have the expertise. In areas where they have the expertise, they share it with us. It works out really well.''

Asked his opinion of the sale, Bennie Williams, the quality control technician at the plant, said, ''It never dawned on me that it was a sale. We've been associated with PCCA since day one and to me it was just like . . . you know, when a woman gets married, she just changes her last name. That's the way it was to me. We just went from ACG to PCCA mill division. The checks still look the same . . . we've still got the same bosses. . . .''

Looking Ahead

The person most instrumental in the sale of ACG to PCCA was C. L. Boggs. He talks about the meaning of it. "One major difference in what we have now is that American Cotton Growers was closed to members except as they came through these 26 or 27 gins, and PCCA is open to anybody who wants to come into the cooperative program." Reflecting the uncertainty and conflict over the pool policy that raged for years, he adds, "I'm not saying that either concept is better than the other; I'm just pointing out the contrast. I'm so excited about what this does for PCCA," Boggs continues. "PCCA has a history of erratic profits, particularly on the marketing side. The warehousing divisions have always been profitable. But markets are either feast or famine. The whole operation is highly volume sensitive," he adds. "If you look at our track record, we do better when we have big crops and big volumes . . . but the denim mill is not so volume sensitive, because all that denim mill needs to make maximum profits is 90,000 bales of cotton. And we can get 90,000 bales of cotton out of a disastrous crop."

What does the future hold? "Out of everything we do," Boggs says, "I still consider PCCA's sole purpose is to market cotton for the farmers we serve. That is all we do even today. Now we market cotton in different ways . . . by selling it directly to textile mills, by selling it to brokers, by selling it through TELCOT, and through denim. Manufacturing and selling denim is just another way of marketing 90,000 bales of cotton." As for tomorrow? "We are not ruling out the possibility of expanding someday, but that certainly is not near term."

Doyce Middlebrook is optimistic about the future of the denim mill. He joins in the universal praise of Bob Hale and knows as much as anyone Hale's value to the cooperative. But he realizes Hale will not be running the mill indefinitely. "Bob is in good health, and he's still plenty capable of managing the mill," he says, "but one of these days he is going to want to retire because he owes it to himself. I think when that time comes, he'll know it, and I think he'll have the right people in the right spots to make it continue to do just what its's doing." Middlebrook believes that Hale "has got in mind the people who can continue to run this plant when he's not there, and I think he's basically got these people in place."

Tom Tusher, Executive Vice-President and Chief Operating Officer of Levi Strauss & Co., worked with ACG from 1984. He sees a continuation of the close relationship between his company and the Littlefield denim manufacturers under their new ownership. But he also sees challenges ahead. The challenges, he says, relate to "what's happening in the jeans market. The demographics in the jeans market between now and the early to mid-1990s are working against us in that the age group that represents the highest per-capita consumption of jeans is declining as both an absolute and percentage in terms of the population." The challenge for ACG "is going to be the need to be able to produce variations on the fabric, possibly different finishes, different weights, more innovation.

"I know that will create some problems for ACG because they assumed the ability to continue to produce the same fabric for a long period of time and that has really been true up to just the last year when the marketplace started changing. With the demographic changes, there also is a move to more flexibility in finishes, more rapidly changing styling and finishing. "My concern would be that the new organization [PCCA] and its capital structure may not feel they have the flexibility to invest in the kinds of equipment changes to give them a broader range of fabric development options, and that could affect, over time, the relationship. We may get to the point," Tusher adds, "where we can't take all of the capacity on a one-product concept. That may not come to pass, but it could. That is the biggest potential 'exposure' that would exist for both of us."

According to Jack Boyd of FmHA, the Littlefield denim mill was the "crown jewel" of the federal rural development program. It was so "not only for Texas, but for a lot of the national programs as far as size of loan and job creation, which," he reminds, "was the thrust of the program to begin with." Lynn

Futch, head of the FmHA Texas office, agrees. "As it turned out, it was one of the best deals that we ever got into in terms of the number of jobs that it created and what it did for the economy of the area. For the amount of money that was involved, it certainly was a very worthy project."

When it was built, the Littlefield mill was the biggest FmHA program of its type. Boyd says, "It remained the biggest one for a long time," eventually being supplanted in that position by "one out of West Virginia related to the steel industry and a lot of pollution control that got up in the neighborhood of $55 million." Futch says that the success of the program, "in terms of the business itself being successful and in terms of what it did for the economy and creating jobs — doing exactly what it was intended to do — has had considerable effect on the image of the B&I [Business and Industrial] program."

If Roy Davis pulled political strings to get the federal loan guarantee to build the ACG denim mill, the results amply redeemed him. Had the project been a failure, Roy Davis, his son Dan, and the board members and management of the cooperatives, would be judged harshly — as incompetents at best, as scoundrels at worst.

Those willing to replicate the West Texas experience might have trouble doing so today, at least in the same way. FmHA's total guarantee authority nationwide now is about $100 million per year. That is equal to one Littlefield mill, at today's costs, for the entire country. And if that isn't restrictive enough, FmHA now is prohibited by law from placing a guarantee of more than $25 million on any one project, and by its own regulations from placing a guarantee of more than $10 million. Does that mean the country will never see another Littlefield backed by federal funds? Not necessarily. Lynn Futch sees an emphasis on the B&I program again because "of so many problems in small towns and rural areas [and] because of a lot of farmers going out of business."

One of the prior ACG directors thinks PCCA will have to expand. "I think any organization has got to grow or die. That's my personal opinion," he says. "You've got to keep an organization going and growing and in enthusiasm and expansion and all that goes with it, or atrophy sets in and you lose it." Based on the history of the West Texas cotton cooperatives, it is difficult to imagine atrophy setting in. It is quiet now. The denim mill at Littlefield runs, continuing to produce first-quality denim for Levi Strauss & Co. The restructured PCCA Board of Directors meets monthly for discussion and decisions on marketing cotton and making denim. The management and staff of PCCA tend to the administration of a single business entity.

There are no daring ventures planned into uncharted territory, but that does not mean the organization has become moribund. Far from it. The staff focuses on finding better ways to do things. For example, PCCA and the National Cotton Council are leading the way in an effort to eliminate the paper warehouse receipt that has been a part of cotton marketing from the beginning. The paper receipts would be replaced with electronic ones. Although not as exciting or adventurous as entry into a new business, it is a project that will have an enormous impact far beyond the Texas Plains. It is estimated that this change, when perfected, might save the U.S. cotton industry 15 to 30 million dollars annually.

The changes now are subtle. They move more slowly than some of the changes of the past, but they move purposefully and forward. Relationships are as smooth as can be expected in a setting where people and their incomes are involved. Although there are some residual hard feelings about old controversies, there are no new ones. The state is one of evolution rather than revolution.

Like the West Texas weather, this is bound to change from time to time. Tranquility, like turmoil, is a temporary condition in West Texas. It comes to the summer sky between storms. And it comes to the cotton cooperatives between revolutions in technology, process, and thought. Roy, Davis is dead. Dan Davis is no longer part of the cooperatives. But the area is too filled with opportunity and the people with spirit to permit malaise to set in. For now, sound management is taking care of business, preserving and building incrementally on what has been done before. What the next major change will be, where it will come, and who will champion it remain to be seen. But it will come. As enduring as the Great Plains of Texas, it will come.

Index

Index

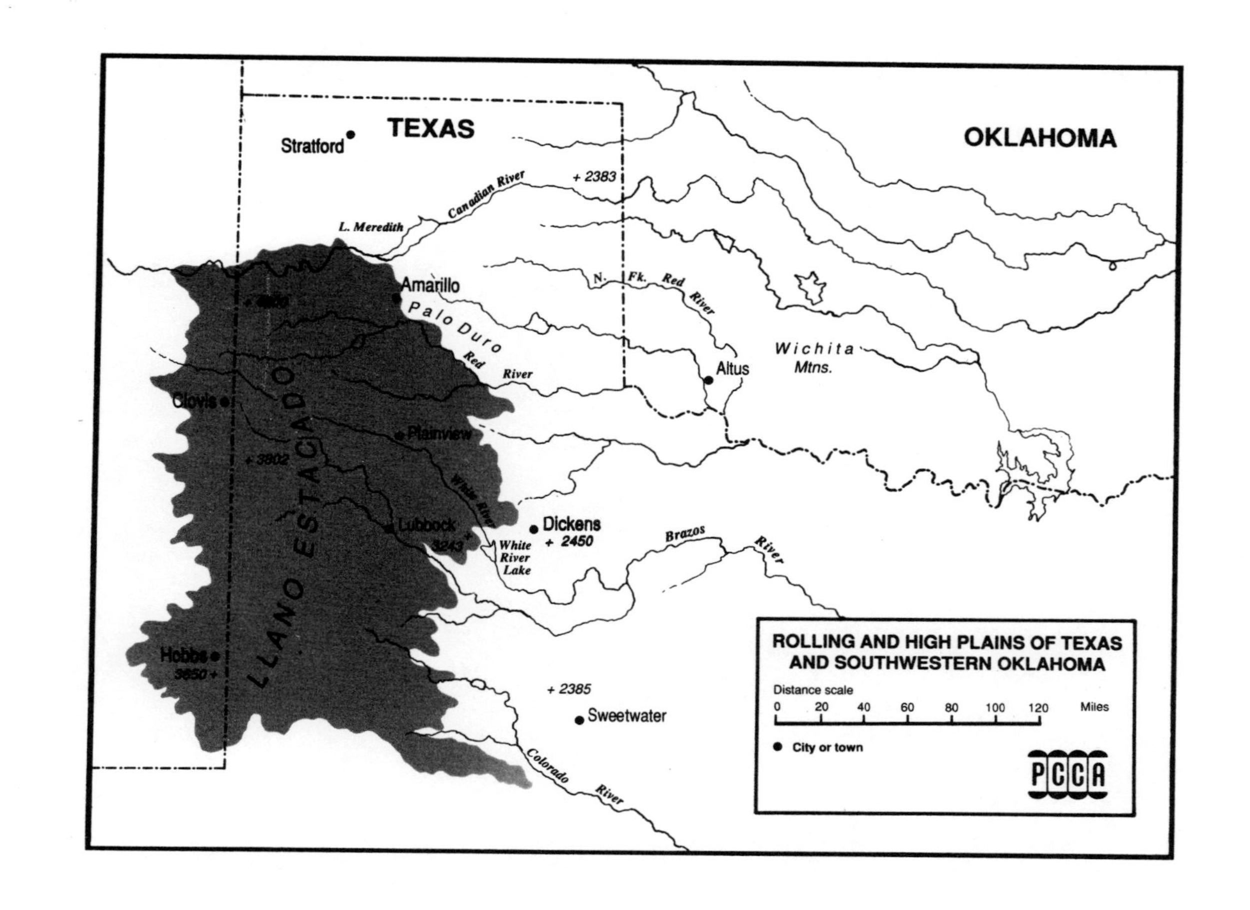
TEXAS
OKLAHOMA
Stratford
Canadian River
+ 2383
L. Meredith
Amarillo
Palo Duro
Red
River
N. Fk. Red River
Altus
Wichita
Mtns.
Clovis
Plainview
+ 3802
White River
Lubbock
3243 +
White
River
Lake
Dickens
+ 2450
Brazos
River
LLANO ESTACADO
Hobbs
3650 +
+ 2385
Sweetwater
Colorado
River
ROLLING AND HIGH PLAINS OF TEXAS
AND SOUTHWESTERN OKLAHOMA
Distance scale
0 20 40 60 80 100 120 Miles
City or town
PCCA

R0161280817
334.
687721
L699

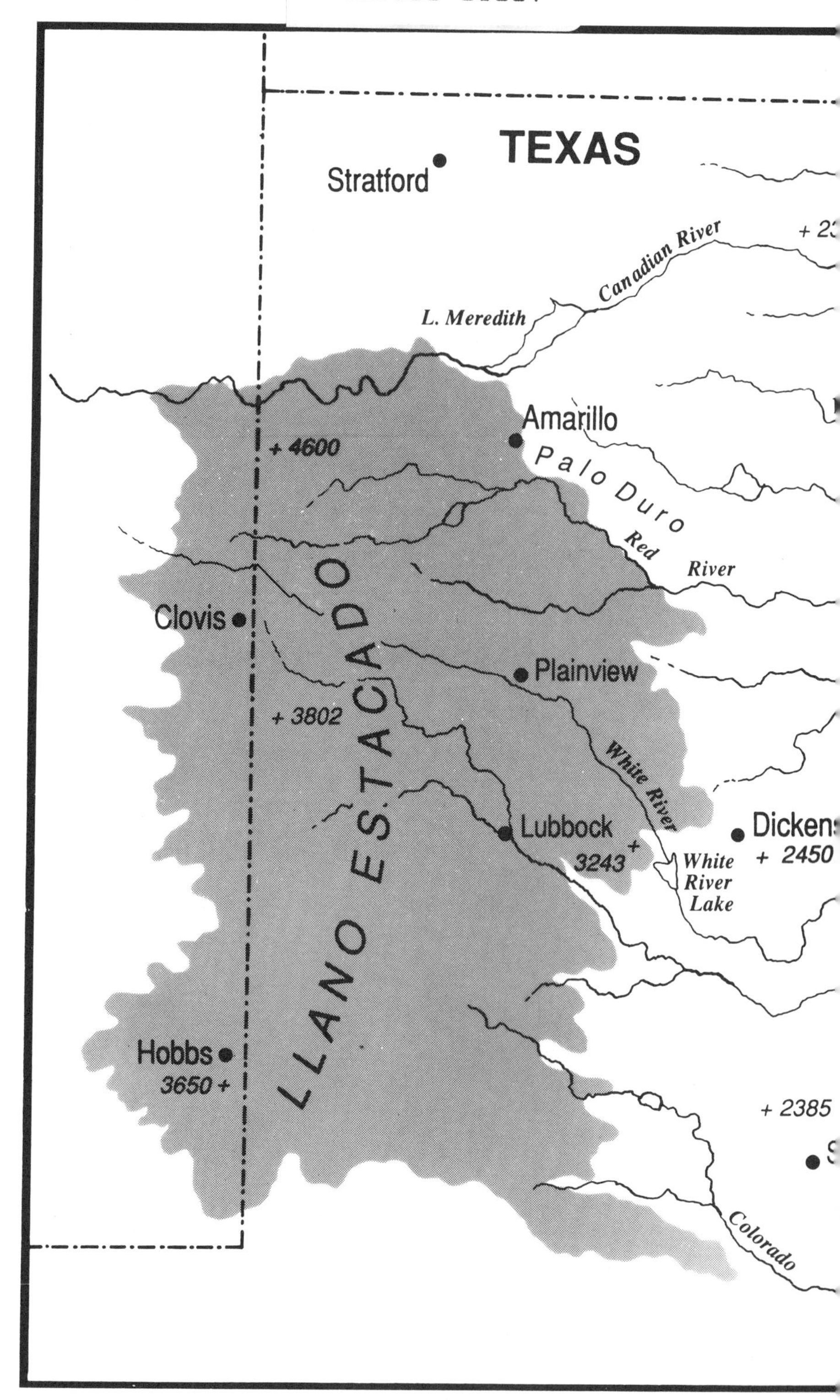
TEXAS
Stratford
Canadian River
L. Meredith
Amarillo
+ 4600
Palo Duro
Red
River
Clovis
Plainview
+ 3802
LLANO ESTACADO
White River
Lubbock
3243 +
+ 2450
White River Lake
Hobbs
3650 +
+ 2385
Colorado